Green Fields Forever

Green Fields Forever

The Conservation Tillage Revolution in America

Charles E. Little

ISLAND PRESS

Washington, D.C. □ *Covelo, California*

ABOUT ISLAND PRESS

Island Press publishes, markets, and distributes the most advanced thinking on the conservation of our natural resources—books about soil, land, water, forests, wildlife, and hazardous and toxic wastes. These books are practical tools used by public officials, business and industry leaders, natural resource managers, and concerned citizens working to solve both local and global resource problems.

Founded in 1978, Island Press reorganized in 1984 to meet the increasing demand for substantive books on all resource-related issues. Island Press publishes and distributes under its own imprint and offers these services to other nonprofit research organizations. To date, Island Press has worked with a large cross section of the environmental community including: The Nature Conservancy, National Audubon Society, Sierra Club, Conservation Foundation/World Wildlife Fund, Environmental Policy Institute, Natural Resources Defense Council, Wilderness Society, National Parks and Conservation Association, and National Wildlife Federation.

Funding to support Island Press is provided by The William H. Donner Foundation, Inc., The Ford Foundation, The George Gund Foundation, The William and Flora Hewlett Foundation, The Joyce Foundation, The Andrew W. Mellon Foundation, Rockefeller Brothers Fund, and The Tides Foundation.

For additional information about Island Press publishing services and a catalog of current and forthcoming titles contact: Island Press, P.O. Box 7, Covelo, CA 95428.

© **1987 Charles E. Little**

All rights reserved. No part of this book may be reproduced in any form or by any means without permission in writing from the publisher: Island Press, Suite 300, 1718 Connecticut Ave., N.W., Washington, D.C. 20009.

Library of Congress Cataloging-in-Publication Data

Little, Charles E.
  Green fields forever.

  Bibliography: p. 147
  Includes index.
  1. Conservation tillage—United States. I. Title.
S604.L57 1986   631.4'5   87-4157
ISBN 0-933280-35-1
ISBN 0-933280-34-3 (pbk.)

Brief parts of this book have appeared in different form in *Country Journal* and in *American Land Forum*.

Photo credits appear following the index.

MANUFACTURED IN THE UNITED STATES OF AMERICA

This book is dedicated to the memory of

EDWARD H. FAULKNER
1886–1964

His agricultural heresy was common sense.
His legacy is the everlasting soil.

I can lime it, cross-plough it, manure it, and treat it with every art known to science, but the land has just plain run out—and now I am putting it into trees in the hope that my great-grandchildren will be able to try raising corn again—just one century from now.

Franklin Delano Roosevelt speaking of his Hyde Park, New York, farm in the mid-1930s

Do you realize that now, for the first time in ten thousand years, people can grow crops without destroying the land?

Greg Schmick, agronomist and executive of a Spokane, Washington, farm-machinery manufacturer, speaking of "no-till" agriculture in the mid-1980s

Contents

Foreword

*No other human occupation opens so wide a field for the profitable
and agreeable combination of labor with cultivated thought as
agriculture. I know nothing so pleasant to the mind as the discovery
of anything that is at once new and valuable—nothing that so
lightens and sweetens toil as the hopeful pursuit of such discovery.
And how vast and how varied a field is agriculture for such
discovery.*

> —ABRAHAM LINCOLN
> (Address before Wisconsin
> Agricultural Society,
> Milwaukee, 1859)

In the United States, agriculture has progressed from an economy
of scarcity to one of abundance in the space of a century. In 1862,
during Lincoln's presidency, Congress passed three pieces of legislation—the act that created the U.S. Department of Agriculture,
the Homestead Act, and the Morrill Land Grant College Act—that
helped the American farmer make invaluable contributions to our
agricultural productivity. If judged only by standards of productivity, American agriculture in the twentieth century has been an
unparalleled success.

At the same time, agriculture in the United States has been marked
by persistent problems, including severe financial stress for many
farmers and serious soil erosion. A wide range of environmental
problems that signal the degradation of natural resources plague
today's highly specialized, capital-intensive agriculture. Production technologies too often have been developed and adopted without sufficient knowledge about their cumulative effects on natural
resources and the environment. Policies designed to conserve soil
and protect the environment frequently have been afterthoughts to

the rapidly evolving crop production methods designed to maximize short-term profits.

National Resource Inventories conducted by the U.S. Department of Agriculture in 1977 and 1982 documented that nearly 1 acre out of every 4 of our 421 million acres of cropland experienced serious soil erosion. The Food Security Act of 1985 recognized the need for new provisions to better link commodity and conservation policy. The Conservation Reserve Program is aimed at converting the most marginal cropland to grass or trees by 1991. By denying federal subsidies, the "Sodbuster" and "Conservation Compliance" provisions should slow down the use of fragile soils for crops. However, since most of the land currently under cultivation in the United States will continue to be needed for the production of food and fiber, erosion will continue to be a major concern.

Conservation tillage is emerging as the common denominator in efforts by people from various disciplines to find a solution to this chronic problem of soil erosion. The enthusiasm of the participants in the first National Conservation Tillage Conference, held in Nashville in October of 1984, indicated the depth and scope of interest in this technique. Few topics continue to spark as much discussion in both agricultural and conservation circles as conservation tillage.

Growing new crops in old crop residue is viewed by conservationists and farmers alike as the most cost-effective means available to achieve soil and water conservation. Many advocates also note that labor and fuel savings are major reasons for the widespread and growing use of conservation-tillage techniques. However, environmentalists are increasingly concerned about the continued reliance on herbicides that characterizes most of the reduced-tillage management systems now being practiced.

Green Fields Forever: The Conservation Tillage Revolution in America is a timely contribution to this discussion. Charles E. Little, a noted author who has written widely on issues of wise land use and agriculture, takes a broad look at both the practice and the promise of conservation tillage. This is an affirmative and provocative book that tells the story of an evolving technology that holds great potential for ameliorating a number of agriculture's problems.

Conservation tillage brings a more complex soil-plant ecosystem back to farming. This complexity, in turn, may provide more balance to nutrient cycling and natural pest management, leading eventually to the discovery of farming methods even less dependent on chemical fertilizers and pesticides. Present knowledge will need to be reinforced with more research to realize fully this promise, which has as its goal a return to health for America's seriously degraded cropland.

In every region of this country's vast and varied landscape, land users are seeking answers to their conservation problems. Innovations developed by the individual farmer, working within his or her local social system, often are more readily accepted than are solutions imposed from the outside. *Green Fields Forever* captures this grassroots approach as an effective channel for technology development and adoption. Through interviews, research, and on-site visits to several farmers in widely differing environments, Chuck Little shows that conservation tillage is an evolving technology that is being shaped by the inventiveness of individual farmers.

The farmer who uses conservation tillage must abandon the traditional system of cropping and learn a set of management skills that require new methods and faith. Conservation tillage is not a panacea, but it is proving to be one of the best ways now known to meet our national priorities of soil and water conservation. Where conservation tillage alone will not control erosion, we can combine it with other practices to substantially reduce topsoil destruction.

Unlike other information available on conservation tillage, which usually focuses rather narrowly on the technical details, *Green Fields Forever* places the technology in a broad context. Beginning with the character of the soil (conservation tillage must work with the soil, letting the soil handle the task of growing plants as it was meant to do) and ending with a discussion of the impact of conservation tillage on the future of American agriculture, Chuck Little has made the story both entertaining and significant.

What is new in this book for you? For those of you who are now concerned with conservation tillage, as environmentalists, conservationists, policymakers, or plain citizens, *Green Fields Forever* offers reinforcement, encouragement, and new information. For those of you who are involved with or interested in America's largest industry—agriculture—this book should enlist your support. A renewed emphasis on conservation tillage is most appropriate at a time when the federal deficit threatens the future of many domestic programs, including conservation research, extension programs, and financial and technical assistance for land users.

When I returned to Idaho after Marine Corps duty in World War II, the Soil Conservation Service and the Soil and Water Conservation Districts utilized every means possible to increase the use of crop residues. Our first task was to stop the use of the match. At that time, the autumn skies darkened with smoke from the burning of wheat stubble as farmers prepared to plow and plant the new crop. Eventually, stubble-mulch farming and crop residue management became part of the technician's conservation tool kit, but the odds of widespread acceptance were stacked against us then.

We have come far in four decades with research, innovation, and farmer participation in soil and water conservation. Now I find it difficult to be anything but a very realistic optimist for the future. Although we need to continue to guard against solutions that create more problems, Charles Little has reinforced my belief that most problems do have solutions.

Green Fields Forever expands our thinking and enhances our understanding. In these pages, you will appreciate anew the dedication of those who till, or no-till, the soil as a way of life for themselves and as the substance of life for you and me.

NORMAN A. BERG
Chief, Soil Conservation Service,
1979–1982

Acknowledgments

This book is a work of synthesis and therefore depended on many, many people for its completion. For helping to organize my research afield—as well as for tutoring me in the fundamentals of agronomy—I owe much to William Moldenhauer at Purdue University, Robert Papendick at Washington State University, and Charles Elkins at Auburn University, all of the USDA's Agricultural Research Service. For generous hours taking me about their own operations and offering endless cups of coffee, pastries, lunches, and dinners, I wish to thank the farmers who are the centerpiece of the book: Todd Greenstone of Brookeville, Maryland; Carl and Rosemary Eppley of Wabash, Indiana; Morton Swanson of Colfax, Washington; and Jerrell Harden of Banks, Alabama. And, while I am at it, I should also like to thank People Express Airlines, without whose cut-rate fares this writer could never have undertaken as much research as he did.

Kenneth A. Cook, an agricultural policy analyst; Roy M. Gray, an agricultural economist; and William Moldenhauer, an agricultural soils scientist, read the manuscript for factual error, misapprehension, and plain wrong-headedness. To the extent that these problems persist, it is my fault, not theirs. I am grateful for the time and care they took with the work. I am also grateful to Sara Ebenreck, formerly editor for the Conservation Tillage Information Center and now editor of *American Land Forum*, for guidance throughout; to Marcella Kogan, for collecting many hundreds of photographs from which to choose the illustrations for the book; to Marvin Craig and Trap Henderson, for their keen literary insights; to Ila Dawson Little, professor of English, whose editorial counsel was essential; and to Roberta Pryor, my redoubtable literary agent. Finally, I should like to thank Barbara Dean, editor at Island Press, who kept the project on track and, most notably, was ever ready to defend the right of the author to express views not always consonant with her own.

CHARLES E. LITTLE
Kensington, Maryland

Green Fields
Forever

1

The Tillage Revolution

Since the achievement of our independence, he is the greatest patriot who stops the most gullies.
—PATRICK HENRY

Nature can take centuries to build an inch of topsoil. And it can be washed away in a season. Franklin D. Roosevelt's farm, two hundred acres, had grown championship corn in the mid-nineteenth century. By the time Roosevelt took it over in 1910, most of the topsoil was gone, and yields were down by 50 percent. He never could bring it back.

Nowadays, the failure of the soil can be made up for by chemical fertilizers, irrigation, better hybrids, and pesticides—to a point. But eventually, "run-out" land makes run-out farms and a run-out agricultural landscape: the lifeless stretches of near-desert in the High Plains and parts of the Far West; the gullied, red-raw lands of the South and parts of the Midwest; the abandoned fields choked with bindweed and puckerbrush in the East, much of it dotted with young red cedar, sure signs that a farmer has left the land.

If Greg Schmick is right—and his view, quoted as an epigraph alongside that of FDR in the front matter of this book, may not be as overblown as it sounds—the fate of farmland need not be the eventual exhaustion of its soils, the buildup of hardpan, or the lowering of water tables. Schmick is young, scarcely past thirty, but he knows about our agricultural history—one of forced migrations of often desperate farmers who could no longer scratch a living from eroded hillsides, undrainable fields, or dried-out land where

Gullies like these in Coffee County, Alabama, as well as less visually dramatic but equally serious soils problems throughout the United States, have plagued American agriculture for centuries. Now, with conservation tillage, there's a means at hand to eliminate virtually all erosion from wind and water and to prevent soil compaction and the "farmed-out" land these problems create.

rain was supposed to "follow the plow" but didn't. He also knows that Americans did not invent agricultural land failure. The two hundred million acres of land we have so severely abused in America is but a small fraction of what has been ruined over the millennia, and is still being ruined, worldwide. Still, we have less excuse than anyone in the world thoughtlessly to deplete what is conceded to be the best patch of farmland on the face of the globe.

Now, perhaps in the nick of time, our agriculture is changing, courtesy of a revolution in the technique of tillage. Curiously, it is a revolution that is going on largely outside of public view. There are no newspaper headlines or television documentaries about it. But it's a revolution nevertheless. It is called "conservation tillage" because it conserves the soil, and the water, and the tilth of the land. When properly practiced, it means that the land need not "run out" as it did for Franklin Roosevelt and millions of others

Economic stability and permanence, as exemplified by this peaceful winter scene of a Connecticut farm, have been elusive goals in American agriculture. Conservation tillage has important implications in this regard. To see it merely in technical terms is to miss the broader significance of this advance in the practice of agriculture. Conservation tillage can help stabilize both the ecology and the economy of rural America, and for that reason it should be understood by all Americans, not just farmers and agricultural specialists.

and that there can be green fields forever, not just for a few generations.

This book aims to tell the story of conservation tillage and how this technical revolution is likely to affect not only farms, and farmers, and farmlands, but all Americans. The plan of the book is relatively straightforward. The next chapter, Chapter 2, introduces, via a young Maryland farmer named Todd Greenstone, one particular technique of conservation tillage popular in the Middle Atlantic states. The chapter goes on to provide a general summary of various other forms of conservation tillage (they are quite different from one another) along with an introductory discussion of benefits, drawbacks, and origins. Chapter 3 provides a historical perspective through a discussion of the tillage technique now being replaced: the use of the moldboard plow. Chapters 4, 5, and 6 show the different techniques in situ in three major commodity-producing regions of the country: the western drylands, the Corn Belt, and the southeastern coastal plain. Chapter 7 takes up the ecology of conservation tillage and the environmental controversy surrounding it: it is a serious one, but it is not insoluble. And Chapter 8 provides some concluding speculation about the effect of conservation tillage on the overall structure and economy of agriculture in the United States.

This is an affirmative book. The tillage revolution bids fair to eliminate gullies and a whole lot more that is troubling our agriculture. And these days, there are troubles aplenty. There are economic troubles, environmental troubles, and most troubling of all, the real possibility of a *cultural* failure in an enterprise—independent farming—that is the historic and symbolic cornerstone of the American way of life.

These are not small matters. And because conservation tillage can affect them in profound ways, this new kind of farming is herein interpreted in a broad context rather than a narrow, mechanistic one—which is the way it has been treated so far in the farm press as well as the few nonspecialist publications that have chosen to discuss it. It's not that they have got it wrong; they have just not got all of it. The agricultural establishment has tended to regard this new concept mainly in terms of chemicals and machinery, not ecology and history. And conservation tillage has to do with these larger issues as well. It is a technology that ramifies into all aspects of our daily lives. And it comes at a time when great changes in American agriculture must take place, changes that we will all be affected by and for which, therefore, we have a political responsibility as citizens.

It's a pleasure, after a long spell of not being able to do so, to write something hopeful about farms and farming in the United States. Conservation tillage may, in fact, be the most hopeful thing to happen to American agriculture since a friendly Wampanoag Indian—his name is lost to history—taught the first colonists at Plymouth how to plant corn in 1621. Later that season, they celebrated, calling it Thanksgiving.

2

A Different Kind of Farmer

The plowers plowed upon my back:
And made long furrows.
—THE BOOK OF COMMON PRAYER

Todd Greenstone is a professional farmer. With his wife, Lois, and two preschool children, he lives and farms in the central Maryland countryside. This part of rural Maryland is charming, a pastoral, rolling countryside favored by wealthy landowners and horse breeders as well as real farmers like Greenstone. He is there because the land is productive and can be rented without difficulty from absentee owners and because he just likes the look and feel of it. He did not inherit a farm; he is not even from a farm family. He is from a family of scientists and teachers. He became a farmer because he wanted to. He remains a farmer because he is good at it, makes a good living, and is his own boss. He is one of a new breed of agriculturist; there is no doubt about that. But not entirely for the foregoing reasons, although they are not irrelevant. He is also different because he does not plow his land. Not in the fall. Not in the spring. He does not turn the soil. His fields are innocent of the furrows of the plow, which has been the primary implement of agriculture for at least six thousand years.

In America, the traditional steel moldboard plow, which John Deere figured out how to mass produce in 1837, became something of a patriotic symbol, for it could cut the tough prairie sod and give the new nation agricultural dominance throughout the world. To plow the land has provided a whole vocabulary of moral metaphor,

Todd Greenstone, a young Maryland farmer, grows corn like this—dense, heavy, fetching top prices. Conventional tillage usually involves the use of a moldboard plow. But Greenstone does better *without* plowing his field, using a version of conservation tillage called "no-till." He simply does not turn the soil. Ever.

the long, straight furrow standing for the long, straight life. "Plow deep while sluggards sleep," advised Ben Franklin's Poor Richard. Never mind. The moldboard plow is not for Todd Greenstone. He does not "till" the soil in any conventional sense. Indeed, his fields may look unkempt to the traditionalist, filled with the stubble and "trash" of prior seasons, for he has substituted conservation tillage for the plow. And, not coincidentally, Greenstone is one of the top producers in his county.

"No-till"—the kind of conservation tillage Todd Greenstone mainly practices (there are many variations, as we shall see)—means what it says: no plowing, disking, harrowing, or other ways of disturbing the soil. ("Tillage," according to the Oxford English Dictionary, includes all forms of cultivation.) The technique is taught at the University of Maryland, from which Greenstone graduated a decade ago. Eventually, he would like to own at least some of the land he farms, but for the moment, he "no-tills" 350 acres of land leased from absentee owners and does custom work on 400 more. Custom

work is strictly a professional service arrangement with estate owners, retired farmers or semi-farmers, and others, such as housing
developers not yet ready to build. But it spreads the cost of a crew
and equipment over a larger base of operations and permits the
quite sophisticated farming techniques, involving costly chemicals
and state-of-the-art machinery, used for no-till.

"What you basically need for no-till," says Greenstone, "is a corn
planter or grain drill, a sprayer for herbicides and pesticides, and
a tractor. And the tractor can be lighter than with conventional
tillage because you don't need as much horsepower to pull a grain
drill or planter as you do a moldboard plow." There is, in fact, a
rusty "six-bottom" plow (six plowshares affixed to a drawbar) in
Greenstone's equipment yard. "There's a use for the moldboard
plow still," Todd Greenstone says. "But not much." And it is in this
way that he parts company with the traditional farmer, who for
generations out of mind has turned the soil with a moldboard plow,
which buried weeds and exposed a fresh surface with which to
make a clean seedbed. Greenstone says that when he has to get rid
of weeds mechanically, which is rarely, he uses a chisel plow, not a
moldboard. The curved steel fingers of his chisel plow do not turn
the soil, exposing its underside to erosion and burying its organic
content, but are good for uprooting pesky perennial weeds that can
get established in an unplowed field after a few years of no-till
despite the use of herbicides to kill them. Sometimes, Greenstone
explains, it is cheaper to get rid of weeds that way than with
chemicals.

But mainly he doesn't chisel, either literally or figuratively, on
his no-till regime, but plants right into the "trash"—stubble from
previously harvested crops along with dead weeds. How can you
plant seed without preparing a seedbed? How can you get the seeds
through the "trash" and into the ground? Pulling a damp tarpaulin
off a sizable and complex piece of machinery, Greenstone shows his
visitor a no-till corn planter, manufactured in fine irony by none
other than Deere & Company. Greenstone is proud of this machine
and pleased that he could get it, used, for $4,500. "It costs $15,000
new," he says. The planter is in effect a ganged set of six separate
units, each with a seed hopper on top that delivers the seed-corn
kernels into a row sliced two or three inches deep by a "no-till
coulter," which is a sharp steel disk that cuts through the organic
residue from the previous season's crop and makes a narrow slit
into the unplowed earth. Set just behind the coulter is a "seed opener,"
two disks slightly offset to wedge open the soil, however briefly.
And, just behind *that*, where the kernel drops into the widened slice,
is a plastic tube that delivers a thin trickle of liquid fertilizer. Fol-

Conservation-tillage farmers in the Middle Atlantic states often use a no-till planter like this, which cuts a narrow slice through the stubble into which seed is dropped from the hoppers, which are followed by a "press wheel" to complete the operation. As the photograph demonstrates, planting can be done the same day as harvesting, permitting farmers to "double crop." This farmer can take two harvests from his land in the same year—barley and soybeans—because of conservation tillage.

lowing is a "press wheel" of solid rubber that closes up the narrow slice as the planter moves by, fast as a man can walk. Such a planter is designed especially to be used on a field with shin-high corn stubble, which bothers it not at all. "Once you get the surface of a field smooth," Greenstone says, "you can double-crop with rotations for several years without replowing." He means that more than one crop per season can be grown if the crops are rotated with other crops. For a no-till field in Greenstone's area, the University of Maryland recommends a two-year rotation starting with corn, followed by a small grain such as wheat, barley, or oats, followed by soybeans, which are double-cropped after the grain with two plantings in the same season.

It seems to work. Greenstone has one field that has not been plowed—not even with a chisel plow—for six years. The corn grows as high in this field as any in the district, maybe higher, and the ears are as heavy as a lead pipe, not from moisture, but from density. Greenstone gets top prices for this corn at the Baltimore market.

Though a description of a no-till planter like Greenstone's makes it out to be a Rube Goldberg device, it is really only an elaboration of the technique used ten millennia ago when Homo sapiens first sliced through the savanna grassland with a sharp stick, dropped seeds into the narrow slit he had made, and pressed it closed with his bare foot. The main difference is chemicals, especially herbicides.

Greenstone uses two kinds of herbicides: a "contact killer," which is sprayed on the seedbed to "burn down" the weeds that emerged with (or after) the previous crop, and a preemergent "selective" herbicide that kills off weeds in the soil before they have a chance fully to germinate. "Paraquat and Bicep," Greenstone explains. "Everybody's high on Paraquat and Bicep." He assures his visitor that unlike the first generation of agricultural chemicals produced during or just after World War II, many of the newer herbicides are made to be nonresidual, breaking down rapidly in the environment rather than maintaining potency over a long period. He also uses new insecticides. "You have to use more than usual, especially at first when you start no-till," he says. The reason is that the stubble harbors more pests than would a clean-tilled field.

Greenstone states (somewhat argumentatively because there are environmentalists in his county who disagree) that the pesticides he uses—the herbicides, insecticides, and fungicides—are safe, "as long as you use the correct rates" in application. Accordingly, he can reduce the number of cultivation trips across a field from six or seven to two. As a result, the organic material remaining on the surface reduces water runoff, so the pesticides (as well as fertilizers and soil) don't run off so much either, showing up in streams and lakes. This is something to consider, in Greenstone's fields especially, for several of them drain into a nearby water-supply reservoir. As for the herbicides seeping down into the groundwater, he says that they tend to be neutralized on contact with the soil, becoming "bound up" by clay particles so that they do not move through the soil but stay in place. Anyway, says Greenstone, you don't use much more herbicide for no-till than you would with ordinary tillage. "In fact, if you do proper crop rotation, you can *reduce* the amount of chemicals used," he says.

Whether a bit more or a bit less, as a practical matter, no-till substitutes the use of herbicides for mechanical cultivation almost entirely. And so, Todd Greenstone, father of two young children, has given the chemical problem considerable thought. "I am concerned about the safety of the new herbicides," he tells his visitor, back in

"No-till" as practiced by Greenstone requires the use of powerful herbicides to "burn down" the previous crop and to reduce crop competition from weeds, which are not removed mechanically via cultivation as with conventional tillage. Here, a no-till field is being sprayed with paraquat, a controversial herbicide in wide use for over twenty years. Some environmentalists believe that many of the older herbicides are more persistent in the environment than their makers and users suggest. Nevertheless, Greenstone believes that chemicals are safe "if you use them safely." And as the father of two small children, he does.

the farmhouse pantry, which he has converted to an office. "But I am not going to worry needlessly about it. I make a living as a farmer, and I do well only if my crops do well. The chemicals are safe if you use them safely."

A corn production work sheet for a typical year provides a list of the pesticides he uses and their cost. They are: Bicep, Bladex, Paraquat, Furadan, and Toxaphene. The first three are herbicides, the other two insecticides. Toxaphene has subsequently been withdrawn from use in Maryland, but farmers may continue to use existing supplies. Taken together, they cost $68 per acre and amount to 26 percent of his projected cost of production, not an unusually high proportion for this area, whether under conventional tillage or no-till. Greenstone's corn-yield goal is 120 bushels per acre. He can often exceed it by a fair amount. He breaks even at 94 bushels per acre.

□   □   □

"No-till" such as Todd Greenstone practices is one of a number of innovative cultivating techniques that come under the rubric "conservation tillage." The variations share a fundamental principle: to leave a substantial amount of crop residue on the surface of the ground rather than turning it under with a plow. Those who are trying to standardize the definition of conservation tillage say that 30 percent residue is a practical minimum. Less than that, and it's "clean tillage" or something close to it, and cannot qualify for cost-sharing payments sometimes made available by the U.S.

Department of Agriculture to establish conservation tillage on a farm.

Given this definition, conservation tillage, according to a 1985 survey,[1] is in place on nearly three out of every ten acres of cropland in the United States—totaling nearly one hundred million acres. Deputy Secretary of Agriculture Peter C. Myers says that conservation tillage will be in use on virtually all cropland within twenty-five years. No new farming technique has been adopted so swiftly in the history of agriculture. The reason is simple: an astonishing array of new herbicides—over one hundred new compounds have been introduced since 2,4-D came out after World War II—are now available to American farmers that are almost directly substitutable for cultivation, three-quarters of which is done solely to keep down weeds. Instead of plowing under the stubble and weeds in a field after harvest in the fall, under conservation tillage it is usually left alone. In spring, instead of another plowing, a "burn-down" herbicide is often used, sometimes mixed with other preemergent herbicides (i.e., herbicides sprayed before the emergence of first shoots of the crop that has been planted). Planting can take place shortly afterward, directly into the residue of last season's crop and the dead weeds. Later in the season, instead of cultivating out the weeds between the crop rows two or three times, further sprayings take place with postemergent compounds carefully placed so as not to harm the crop.

The results are impressive. Yields are usually no less, and sometimes more, than with conventional "turn-the-soil" tillage. Passes over the field by heavy tractors are reduced to a minimum, which in turn reduces costs (of tractors, tractor fuel, maintenance, and the man-hours to run them) and reduces the compaction of the soil as well. A farmer can, as Todd Greenstone says, spend more of his time riding the desk than riding a tractor—and that's where the profits are made, at the desk.

In areas of the country where Greenstone's kind of no-till is not practical, other conservation-tillage strategies have been devised. In the Midwest, where drainage is often poor and the soil takes a long time to warm up in the spring, "ridge-till" has taken hold. The technique is to create permanent, raised planting beds—ridges— the tops of which are cleared of weeds and residue and planted. The residue remains on the field, however, between the ridges. In

---

[1] The 1985 Survey of Conservation Tillage Practices is published by the Conservation Tillage Information Center, a national clearinghouse for information on conservation tillage. The center is administered as a special project of the National Association of Conservation Districts.

this version, somewhat more mechanical weed control is used compared with no-till, which relies entirely on chemicals. In some areas of the Great Plains and in the South, "strip-till" is coming into favor, in which only a small percentage of the land is cultivated: a strip to make a seedbed. In the western dryland farm areas, a form of no-till is used but with such different machinery that it seems unrelated to the kind practiced by Todd Greenstone.

In addition to these quite new forms of tillage, many farmers have given up the moldboard plow but not the chisel plow and pretty much substitute the latter for the former in conservation-tillage regimes, sometimes with astonishing results in yield. The "champion" corn grower in the United States uses this technique. While some chisel plows are virtually indistinguishable from moldboards in terms of burying organic matter, others disturb the surface of the field very little, leaving the bulk of the residue in place, nor do they plow very deeply in any case. Finally, there are additional variations in tillage techniques—experts have created a catchall category for them called "reduced tillage"—that can leave a significant amount of organic residue from stubble and dead weeds on, or just under, the surface of the soil.

All the emphasis on residue is of more than academic interest to nonfarmers, for it is the presence of residue—organic matter on or near the surface of the soil—that can substantially check soil erosion and, by retaining moisture near the surface, can reduce the need for irrigation in certain areas of the country. (One farmer in water-shy Nebraska says that no-till in any given growing season is equivalent to a good, soaking, two-inch rain.) In the western and southern plains, the tough native grasses have long since been plowed under. Without irrigation to grow a crop and therefore hold the dry, powdery soil in place, millions of acres are subject to severe wind erosion and eventual desertification. In moister climates, even more millions of acres of topsoil are affected by sheet and rill erosion. Sheet erosion is the general washing away of soil without much visual indication of it. In rill erosion, small rivulets—rills—are cut into the soil. Sheet and rill erosion have already stripped the topsoil from the New England hill farms, the Ozarks, most of the southern Piedmont, and along substantial parts of the Mississippi drainage. The soil is carried down creeks and into rivers to make shoals or to the seacoast to fill shipping channels and even create new estuarine landforms. Such widespread effects of soil and water loss through traditional agriculture is what makes conservation tillage appealing not only to the farmer but to the general public as well, for the public welfare is clearly a beneficiary of it.

□   □   □

The green fields where conservation tillage is practiced stand in vivid contrast to the Dust Bowl days of the 1930s. This photograph was taken in Cimmaron County, Oklahoma, in April 1936. Although scenes like this were never to be repeated, "little Dust Bowls" have returned to the Plains—in the 1950s and again in the 1970s. The recurrence has given conservation tillage a powerful impetus.

Why has erosion gotten the upper hand? A bit of history might help. World War I, like most wars, drove up the price of commodities so that farming, in those days, could produce a better livelihood and life-style than farmers had been used to before. But then, during the 1920s, agriculture, much expanded, went into a price slump from which it was not to recover until another war twenty years later. So farmers of the 1920s, using the new tractors that were now affordable, took farming into new areas never before put to the plow. More and more new lands were opened in farmers' frightened efforts to maintain income in the face of falling prices. Then came the Wall Street Crash of 1929 and the beginning of a mounting depression in all sectors of the economy, not only in agriculture. So the farmers drove themselves even harder. This desperate strategy seemed to work for a time, for 1929 and 1930 were wet years and the crops grew well, making up, at least in part, for income lost through falling prices.

The first warnings came in 1931 and 1932 on the Great Plains, where small dust storms began lifting the thin layer of topsoil from the plowed-up, dry prairie land in Kansas and eastern Colorado. What began in a small part of those two states became pandemic by 1935. On April 14 of that year, a young drifter, balladeer, and political activist from Pampa, Texas, watched as a black, airborne tidal wave swept down from the north. It consisted of thousands upon thousands of tons of dust from farms as far away as the Dakotas. Visibility was near zero. Animals died of asphyxiation, as did babies and small children whose mouths and noses became clogged with dust—somebody's topsoil. Humidity dropped to a parching 10 percent. In a panic, people fled indoors, trying to chink the windows and doorjambs with wet rags in the whistling, gritty gloom, as the wind-borne dust sought entrance at every fissure, every crack, in a kind of demon retribution. When it was over, spring plantings had been torn out by the roots or inundated with drifts of dust. Dead animals were partially covered by dust dunes. In the houses, there were dark corners that would never be free of dust in their entire existence. And this was only one storm; there'd been worse ones before and there might be worse ones to come. So Woody Guthrie wrote his famous song "So Long, It's Been Good to Know Ya" and, like tens of thousands of others, picked up and left the plains. A long, strung-out column of flivvers stuffed with furniture and children abandoned the land they had plowed and plowed, never to return.

When it was all over, the dust storms only a memory, one hundred million acres of American farmland had been ravaged. This acreage could be added to another hundred million acres of gullied farms in New England, the Appalachians, and the Ozarks and hard-panned fields with failing crops in a sizable fraction of the flat or gently rolling landscape of the Midwest and South.

Into this crisis came Hugh Hammond Bennett, a North Carolina farmer's son who was to become the father of soil conservation. Big Hugh, as he was called, or sometimes "the Chief," was appointed by President Roosevelt in 1935 to create the U.S. Soil Conservation Service. Bennett lobbied Congress hard for laws and money for his programs. During one crucial hearing in Washington, the legislative chamber was darkened as a dust cloud passed, driven eastward by the prevailing winds. "There goes Oklahoma," said Bennett, and he got what he came after—the wherewithal to help farmers fasten down their soil.

For a while, it seemed as if Hugh Bennett's new soil-conservation program might work. Technical assistance and financial support to soil- and water-conservation district work—and the districts were set up in virtually all agricultural counties—produced the shelter belts, terracing, contour plowing, strip cropping, cover crops, grassed waterways, and other techniques to hold topsoil in place that Bennett and the soil scientists before and after him developed. Slowly

During the 1930s, Hugh Hammond Bennett, the "father" of soil conservation, introduced new soil-saving conservation measures promoted by the U.S. Soil Conservation Service, such as the contour strip cropping and grassed waterways shown on this Wisconsin farm. By the late 1950s, when this photograph was taken, many had declared a victory over soil erosion.

at first, then with gathering momentum, the program was put into practice over the next twenty years with measurable success on farms ranging in size from thousand-acre spreads to the old forty-acres-and-a-mule clearings that remained on southern hillsides. In Iowa, according to John F. Timmons, an economist at Iowa State University, soil loss in the erosion-prone corn lands of the western part of that state dropped from twenty-one tons per acre in 1949 to fourteen tons per acre in 1957 because of conservation practices. At that rate, it seemed to many that some kind of equilibrium might well be reached as the programs of the Soil Conservation Service took hold. Many declared a victory over erosion, and citizen support groups devoted to promoting soil and water conservation—such as Friends of the Land, organized in 1940 by prominent businessmen, scientists, and civic leaders—passed into memory. Agricultural surpluses mounted along with yields per acre, which had increased by 50 percent in the decades since the end of the war. President Eisenhower (at one time a director of Friends of the Land) established policies that took acreage out of production, for the old Liberty ships, riding at anchor in an endless row along the west bank of the Hudson River at Haverstaw Bay, were already filled with rotting surplus grain for which there was no market. The surplus was so astonishing, and there was so little need to plant marginal lands, that finally, the work of people like Big Hugh Bennett, novelist Louis Bromfield, and scientist Paul Sears, who had led the conservation movement in the 1930s and 1940s, was all but forgotten. Erosion became a textbook subject and a wholly technical matter—of concern to farmers, of course, but scarcely a public issue.

The 1950s and 1960s brought affluence and a new spurt of growth in cities and suburbs that drew attention away from the countryside. Even a "little dust bowl" in the early 1950s went largely unmarked except by those who had to experience it. Then in the mid-seventies, the dust devils started kicking up again in Colorado and Kansas. While intermittent dust storms were not unusual, this event presaged something different. In a way, it was reminiscent of the 1930s, for a second great "plow-up" of marginal lands had begun.

Just two years after the President's Council on Environmental Quality had (in 1975) assured Americans that "a new dust bowl is unlikely," a good many farmers in the West and Midwest were hard pressed to tell the difference between the mid-thirties and the mid-seventies. In one Kansas town, street lights had to be turned on at 10:00 A.M. when visibility approached zero because of the dust. Paul Wilcoxen, a wheat farmer in Hamilton County, Kansas, which stood near the epicenter of the worst of the dust storms of the 1930s, lost 40 percent of his crop, according to an account that year in *U.S. News and World Report*. The dust smothered plants, or if not, they were pulled out of the ground by the wind. Even in the Wilcoxen home, dust was everywhere. The baby choked and had to be taken to the doctor. "We could see the dust coming in through the electric plugs in the walls," Wilcoxen said.

Meanwhile, back in Iowa, Professor Timmons saw that the downward trend in erosion statistics had reversed itself. Soil losses from water as well as wind had gone back up from the fourteen tons per acre per year of the late 1950s to over seventeen tons by 1974, and the director of Iowa's Department of Conservation allowed that "a farmer is losing two bushels of topsoil for every bushel of corn he produces." Soon enough, it became clear, as later research revealed, that soil loss had become as serious a matter as it had been during the Dust Bowl thirties.

After the famed wheat deal with Russia in 1973, which greatly increased demand for U.S. commodity crops, marginal land was put to the plow, which in many places wiped out earlier gains in soil conservation made by the Soil Conservation Service. Dust from Texas, where this photograph of a buried wheat crop was taken, began showing up on cars parked in Ohio.

A concatenation of forces, beginning for the most part in the early 1970s, had made this so. One was the famed Russian wheat deal, which produced a large, new market for surplus grains and helped keep prices high. As food prices went up and grain reserves were drawn down, Secretary of Agriculture Earl Butz, forgetting his history, urged farmers to "plant fencerow to fencerow" to cash in on the bonanza. Farmers did him one better: they took the fences out, and along with them the hedgerows and shelter belts, as they bought or rented more and more land to put to the plow. Wasn't it the age of affluence in the cities and suburbs? Why not on the farm? So they mortgaged to the hilt to buy land and to buy bigger and bigger tractors, big as battle tanks, that could create mammoth swaths of turned earth with twelve-bottom plows. They moved rapidly across the land in long passes across joined fields stretching to the horizon. Such an agriculture did not permit small fields bounded by wind-

breaks, or complex contours, or strip cropping, or terraces, or any
of that. Up hill and down dale went the plows without regard to
the pelting rains. In the West, the plows advanced farther and farther into arid land, no longer caring whether rain followed the plow,
as the adage had it, because they had something better than rain:
they had center pivots to make the parched soils of the High Plains
yield crops as if they were in the wet subtropics. "Climate-free agriculture" it was called.

Most soil conservationists knew what was coming all through
the decade of the 1970s, but they had no statistics on a national
scale to prove it. Then came the National Resources Inventory of
1977, which showed that an astounding one-third of U.S. cropland
was eroding at rates well above the capacity of the soil to regen-
erate. At some point, it was obvious, such erosion had to become
significant to the overall productivity of the American land.
According to David Pimentel of Cornell University, a topsoil depth
of ten to twelve inches can produce up to 125 bushels of corn per
acre. When half that topsoil is eroded away, as it is in vast areas of
the country, the maximum yield under similar cultivation methods
drops to 83 bushels per acre.

In the East, Midwest, and South, sheet and rill erosion were
shown to be taking an amount of topsoil—some five billion tons a
year overall—that rivaled even the worst years of the 1930s. In the
High Plains, another problem was developing with the Ogallala
aquifer. The Ogallala is a huge, underground, water-bearing for-
mation of gravel, clay, and caliche, six hundred feet deep in places,
that irrigates a north-south band of cropland stretching from the
South Dakota–Nebraska border down through the Texas panhandle. In 1950, less than seven million acre-feet of water was withdrawn from it for agriculture. But by the 1980s, more than three
times that much was being pumped out of the Ogallala every year
for the center-pivot sprinkler systems which can irrigate a quarter
section—160 acres—at a time, as well as for the older irrigation-
ditch fields. Farmers, agribusinesses, and rural towns and cities
over the Ogallala prospered. The trouble is, the aquifer replenishes
itself at the rate of one-quarter to one-half inch a year from natural
seepage. But the water is pumped out so fast in some places that
farmers have to redrill their wells every year. In some places, the
overdraft is as much as 95 percent. Toward the end of the 1970s,
some wells had already gone dry, some had to be redrilled so often
they were just abandoned, and others consumed too much energy
to make it worthwhile to pump the water out. The experts now
predict that the Ogallala will last another twenty years, maybe
thirty. After that, folks will have to return to dryland farming: not

Center-pivot rigs like this can irrigate a quarter section—160 acres—of land at a time and have permitted former rangeland to be intensively cropped. Much of the water comes from the huge, multistate Ogallala aquifer, running from Nebraska southward to the Texas panhandle. Irrigation can triple or quadruple yields. Yet the aquifer is being so rapidly depleted in places that some ecologists fear a new Dust Bowl if the irrigated land, whose original grassland ecosystem has long since disappeared, goes out of crop agriculture, since crop roots and irrigation water have been the only things holding the soil in place.

so much corn, more wheat, and a good deal less yield per acre for almost everything. Irrigation can triple or quadruple productivity in arid areas. It works the other way around, too. Moreover, it is only the irrigation from the center pivots that has held the soil in place. What will happen, ecologists have begun to wonder, in those places where the center pivots have turned their last? When the crop roots have gone, what roots will there be to keep the soil from blowing away again, as it did in Woody Guthrie's time?

At the very moment that the agricultural establishment was getting this shot of bad news, the all-too-brief and dizzying age of fencerow-to-fencerow agrarian affluence suddenly disappeared. A couple of bad-weather years in a row, a reduction in prices, higher interest rates, plus higher costs for everything else, from diesel fuel to blue denims—all this was beginning to bring ruin to many overextended farmers, particularly young farmers who had borrowed heavily during the boom years. In 1979, they came to Washington, as few people in that city will forget, with their big, dual-wheeled tractors, for a mammoth demonstration and a "plow-up" of the grass on Washington Mall (which they later reseeded) from all their racing around. But it was too late, and anyway they were talking to the wrong secretary of agriculture. Butz was replaced by Bob

Bergland, who had not led the farmers astray in the first place, and who in the second, couldn't do anything about their plight. And, despite the largest subsidies in the history of agriculture,[2] neither could Bergland's successor, John Block, or his replacement, Richard E. Lyng, both appointed by President Reagan. So the farmers went home, and went broke, and got foreclosed, and became part of the tragic statistics of a new agricultural depression. In 1986, the Farmer's Home Administration—the federal government's lender of last resort—sent sixty-five thousand "debt notices" to farmers, the first step in foreclosure proceedings. At the very time when farmers should be reinstituting those conservation practices, which many abandoned during the halcyon days—terracing, contour plowing, strip cropping, and the rest—they are flat out of money to do it, and no American in his right mind either blames farmers or makes demands on them, given the current economic crisis across the countryside.

□   □   □

And there's no place else to go: no "new lands" to put to the plow for the first time. Now, after two centuries of agricultural expansion, the land for crops is pretty well identified in resource terms. It totals about 420 million acres, with about 375 million acres of it in intensive use, including about 50 million acres that are irrigated. What we have now in cropland may be about all we'll ever have. It is this cropland base that now produces, and must continue to produce, most of the food and fiber consumed by ourselves, used to feed animals, and sold or distributed abroad. While additional acreage in pasture and range is necessary to agriculture, even meat products depend largely on the cropland base to provide feed for the pig parlors and cattle feedlots that are increasingly being substituted for pasture and range as a way to produce meat.

And so the mighty enterprise of American agriculture—the largest sector of our economy and still our largest export category—rests, in substantial part, on a land base that, while sizable, is still scarcely more than one-fifth of the land area of the nation. Consequently, its ecological status—and the stability of soil is the sine qua non of ecological viability for the production of crops—is of major importance to everyone. When the soil goes, so goes agriculture, and so goes the historical foundation of the U.S. economy. But that is where Todd Greenstone comes in—a different kind of

---

[2]Farm price supports began to soar in 1982, from $10 billion that year to $25 billion in 1986. During the decade of the 1970s, farm subsidies exceeded $5 billion only once (1978).

The future of agriculture rests on the soil. And conservation tillage, as practiced on this Kansas farm, provides a practical way, for the first time, to preserve soil and soil moisture and yet maintain yields and farm income—perhaps even improve them.

farmer from those who, inadvertently or not, created the current crisis in soil erosion. Here is the message from university scientists and government officials on what the new conservation-tillage farmers can do for the soil of the American cropland base. They say that conservation tillage can cut soil erosion on it, from wind or water, by 90 percent.

Ninety.

One hopes Todd Greenstone has a lot of friends as "different" as he. For with conservation tillage, there would appear to be, for the first time, a practical way to preserve our soil and soil moisture and yet maintain, perhaps even improve, yields and farm income. Conservation tillage relies on the careful orchestration of a congeries of technical advances, including new chemicals, new hybrids, and a new generation of farm-machinery design. To be sure, there are rather more chemicals of uncertain environmental effect than there ought to be. Moreover, some of the new hybrids are coming about through genetic engineering, itself a controversial issue. And as for machinery, some (but by no means all) of it required by conservation tillage costs so dearly that only the very largest farmers can afford it, or can afford to replace their existing machinery.

But those problems, important as they are, take nothing away from the fact that conservation tillage is historically significant as a technological event. Its basic idea, after all, is an agricultural heresy—that to plow a field, to turn its soil, is not the best way to grow a crop, but the worst. Where the moldboard plow turns the soil over, the new tillage keeps the soil "right side up." John Deere's rugged plowshare was necessary to convert the prairie into cropland, the essential realization of the nation's westward expansion. But the continued use of it reduced a great deal of that cropland to a dust-ridden near-desert in its dryer parts and to a vast expanse of overworked, infertile, gullied, and hard-panned fields almost everywhere else.

Only a few knew it at the time, but when in the 1930s the dust boiled up blackly from the High Plains and the dark, sweet loam sluiced down creeks and rivers to make shoal water in the broad, muddy rivers of the Middle West, a generation before Todd Greenstone was born, the age of the moldboard plow was over.

3

The Saga of the Moldboard Plow

*This is one of those times when only the dreamers
will turn out to be the practical men.*
—LEWIS MUMFORD

Surely, if ever there ever were a single, perfect, symbol for the American ethos, it would be the moldboard plow. The virgin American land was made for this plow; manifest destiny was achieved with it; the wealth of the nation depended on it.

In *Pioneer Women,* Johanna Stratton's remarkable collection of contemporary accounts of settling the Kansas prairie a hundred years ago, the importance of the plow is made plain by a farm woman named Mattie Huffman. She writes: "On April 1, 1875, we got to our home near the southern state line. Father had located the place several months before. All the improvements that the place had on it was a meager log house from which the chimney had fallen out. We had barely gotten unloaded when a neighbor man rode up on a donkey and brought us the key to our house. You are no doubt wondering what was locked in the house. Well, it was something very precious—a plow. Father had stored it there after he had located the place. It was to be our means of support."

The plow in Mattie Huffman's house was specially designed for the prairie—a heavy iron "breaker" plow or "prairie breaker"— with a share that could slice a shallow strip of sod twenty or thirty inches wide and with a long, low moldboard that could turn over the furrow slice evenly and perfectly to expose the damp, matted

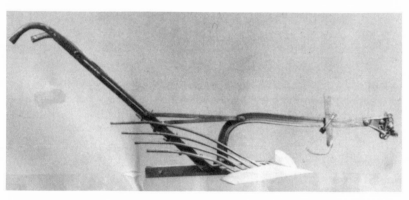

A Kansas "soddy" and the "grasshopper plow." A century ago, the first settlers of the Great Plains broke the sod with a moldboard "prairie breaker" plow such as this and used the turves to build their homes. Ever since, the moldboard plow has symbolized the American pioneer spirit.

roots of native grasses underneath, providing a rudimentary seedbed. As Stratton describes the process, it took several yokes of oxen and sometimes additional manpower to pull the plow through the roots of the perennial prairie grasses, heretofore undisturbed for thousands of years except by the hoofs of thundering buffalo, that made up this complex climax ecosystem. After plowing, according to another pioneer, Mrs. J. H. O'Loughlin, "we . . . used an ax or hatchet to make a hole in the sod, then dropped the seed and closed the hole with our heels."

And so it was that the great heartland was settled, by a single-bottom walking plow and an ax, an acre at a time—an acre, as defined in Biblical times, being the amount of land that could be plowed by a yoke of oxen in one day. An acre a day in Kansas would be very good progress indeed. But the progress was made more quickly than anyone would have thought possible. The Homestead Act, signed by President Lincoln in 1862, transferred the ownership of 147 million acres—in 160-acre parcels—from the federal government to some 1.6 million families like the Huffmans and the O'Loughlins. In scarcely more than half a century, the heartland was conquered. Farm population peaked in 1910 at 32 million and the number of farms peaked just ten years later, totaling 6,454,000. And they all had moldboard plows. It was an agriculture unlike any before. As agricultural historian Walter Ebeling puts it, "the very abundance of land led to a focus on *extensive* rather than European-style intensive agriculture. . . . The farmer was less concerned with how much each acre could yield than with how many acres he could clear and crop." Writing with an enthusiasm typical of the day, the Reverend Josiah Strong, a Congregational clergyman and author of *Our Country*, published in 1886 and ultimately selling 175,000 copies, offered the following analysis of nineteenth-century American agricultural achievement:

> The crops of 1879, after feeding our 50,000,000 inhabitants in 1880, furnished more than 283,000,000 bushels of grain for export. The corn, wheat, oats, barley, rye, buckwheat and potatoes—that is the food crops, were that year produced on 105,097,750 acres, or 164,215 square miles. But that is less than one-ninth of the smallest estimate of our arable lands. If, therefore, it were all brought under the plow, it would feed 450,000,000 and afford 2,554,000,000 bushels of grain for export.

□   □   □

The agency of this abundance, the moldboard plow, was in Josiah Strong's day still a relatively recent invention, no more than 150 years old. For literally ages, the plow—whether of wood, wood cased in iron, or even cast iron—was meant primarily to open the soil rather than to turn a furrow slice. There were many different plow shapes, usually variations on a wedge and drawn through the soil by man or beast or both. But the basic method was to open the soil literally by plowing through it, as a ship might plow through water, then to go over the field again with hand hoes or mattocks (or, later, a drawn harrow) to break up the clods in order to prepare a seedbed.

To make a field perfectly smooth, which was difficult even after the most careful harrowing, there was cross plowing, still a common practice today. Writing in the *Georgics*, published in the year 30 B.C., Virgil counseled: "Plough the fallow in early spring, and plough frequently—twice in winter and twice in summer. . . . Harrow down the clods, level the ridges by cross ploughing."

It took seventeen centuries after Virgil for "high farming"—the foundation of scientific agriculture—to begin to develop in the Western world, beginning in the Lowlands, then moving to England. In 1701, Jethro Tull, an English agriculturist, raised the curtain on the Agrarian Age by introducing the first mechanized seed drill to replace broadcast planting.[1] And later, Tull published the famous treatise, *Horse-Hoe Husbandry*, recommending the repeated cultivation of the soil. He believed that plants grew by taking up minute particles of earth. It therefore followed that the more the soil was pulverized, the better could plants be nourished. Accordingly, Tull advocated frequent plowings of soil plus hoeings to kill the weeds between planted (drilled) ridges. He actually grew thirteen consecutive crops of wheat on the same field without manure using this method.

At about this time—the early 1700s—a new invention from the Netherlands showed up, called the "turning plow." Affixed to one side of the plowshare was a wooden board, tilted over so that when the share lifted the soil, the furrow slice caught on the front edge of the board and rode along it as the plow moved forward, causing the slice to turn on its side, or in some cases, completely over. By the end of the eighteenth century, the turning plow, with a board that molded the furrow slice so that its underside would be exposed and broken down by air, rain, and frost, was in general use in Britain. And a century after that, it had spread across the whole North American continent.

No one considered, of course, that the disintegrating influence of air, rain, and frost might deplete the soil's nutriment, not even that redoubtable American agronomist and agricultural engineer, Thomas Jefferson. While serving as envoy to France in the late 1780s, Jefferson visited Holland and the Rhine Valley to see "high

---

[1]To "drill" seeds does not mean to force them into the ground, as the common usage of the word may imply. "Drill," in this case, is from a different root; the Oxford English Dictionary suggests it may be "rille," a small channel or furrow (rill) formed by a rivulet. Tull himself wrote that he called his invention a drill "because when farmers used to sow their beans and pease into channels or furrows by hand, they called that action drilling." Modern-day planters used for small grains, and sometimes for close-grown soybeans, are also called "drills."

Thomas Jefferson invented an improved moldboard plow for which he won a coveted prize from a French agricultural society. He believed the moldboard plow to be "the most useful of instruments known to man." Even so, he was aware that the misuse of the plow could lead to erosion.

farming" firsthand and became interested in the physics of the moldboard plow. "The offices of the moldboard," he wrote in a memorandum, "are to receive the sod after the share has cut under it, to raise it gradually and reverse it. The fore end of it then should be horizontal to enter under the sod, and the hind end perpendicular to throw it over, the intermediate surface changing gradually from the horizontal to the perpendicular."

Jefferson returned home to be President Washington's secretary of state, but he resigned the post before the completion of his term and returned to his beloved Monticello in Albemarle County, Virginia, in 1793. He owned, at Monticello and in Bedford County to the southeast, about ten thousand acres. At fifty, Jefferson thought he could devote the rest of his days to what he believed was man's noblest pursuit, farming.[2] As it happened, he returned to public life three years later, as vice president to John Adams, and then he was himself elected president for two terms. But in this "blessed

---

[2] Jefferson made a distinction between a "farmer" and a "planter." A farmer was a man of intellect and discernment, a professional with a long-term commitment to the scientific development of agriculture, to the stewardship of agricultural resources, and to the cultural advancement of rural society. The planter was often a parvenu, an exploiter who destroyed resources and added nothing to learning or to the welfare of the community. Tobacco was the planter's principal crop, and the most destructive of all for the soil. Jefferson despised growing tobacco, but occasionally had to resort to it himself when short of money.

interlude" of three years, the Monticello farmer built a working model of a design that significantly improved on the clumsy plows he had seen in Europe. Jefferson called his design the "moldboard of least resistance," for it had a sharp "toe" instead of a blunt one to receive the furrow slice and was curved gracefully so that the pressures of the earth moving along it were evenly distributed. Some years later, Jefferson sent his specifications to a foundry in Richmond (early versions were of wood and could be made without difficulty at Monticello). He wrote of the results to his friend Charles Wilson Peale: "I have lately [1815] had my mouldboard cast in iron, very thin, for a furrow of 9. I. wide & 6. I. deep, and fitted to a plow, so light that two small horses or mules draw it with less labor than I have ever before seen necessary. It does beautiful work and is approved by everyone." Jefferson had every right to be proud, for his contribution was an important one, though he never patented it. In 1805, the design won him a medal from the prestigious French Société d'agriculture du département de la Seine. Though he was by then president of the United States, he was so extraordinarily pleased with the honor (shared, as a matter of fact, with a simple French tiller of the soil who had also improved on the design of the plow) that when the medal failed to arrive, he asked a friend to intercede with the society in his behalf to make sure the award was delivered to him. Later, French agricultural experts wrote him, with typical Gallic certitude, that they had determined his design for the moldboard to be mathematically perfect, and it could not be improved upon.

Given such a history, one does not attack the efficacy of the moldboard plow casually. It would be almost like criticizing the design of the American flag. Indeed, Jefferson, like many farmers of his day (and of this), loved to plow. The plow was, he thought, "like sorcery" and was "the most useful of instruments known to man." In 1785 he kept his slaves afield for 220 days of plowing and complained that the weather had reduced the amount of time they could spend turning the earth. Even so, Jefferson was aware of some of the dangers of misusing the plow and became a promoter of "horizontal plowing," which we now call contour plowing. Jefferson said that his son-in-law, Colonel Thomas Mann Randolph (later governor of Virginia and a U.S. congressman)—"the best farmer in the state"—had developed contour plowing and "has really saved this hilly country." The soil, wrote Jefferson, "was running off into the vallies with every rain, but by this process we now scarcely lose an ounce. . . . The horizontal furrows retain the surplus [water] until it is all soaked up. . . ."

An early John Deere steel moldboard plow. This plow, made in 1838, was a forerunner of the mass-produced model credited with opening the Great Plains to agriculture. It was light, tough, and inexpensive. And it could zip open prairie sod and heavy bottomland clays formerly thought to be unplowable.

But not everyone was so wise. In 1837, a thirty-three-year-old blacksmith from Rutland, Vermont, came to the Illinois prairie to set up a smithy. His name was John Deere, and with a partner, Major Leonard Andrus, he eventually developed a means to produce moldboards of steel inexpensively. By 1850 he was shipping sixteen hundred steel plows a year throughout the country. There is a theory, and it may very well be correct, that John Deere's mass-produced steel moldboard plow (manufactured in Moline, Illinois, where Deere & Company still has its headquarters) changed the history of the nation, for now a plow was light, affordable, and would last. With a stout team, a man—any man, not just a wealthy landowner like Jefferson with a scientific turn of mind—could cut deep into black bottomland soils and zip open prairie sod that had formerly been unplowable. The Deere plow and its imitators meant the opening—and the plowing up—of the vast agricultural heartland of America, "the nation's great endowment," as Walter Ebeling puts it, which lies between Denver, Colorado, and Columbus, Ohio, and runs from border to border—a natural resource that seemed endless and inexhaustible when John Deere's steel plow was introduced but that exactly one hundred years later, during the time of dust and depression, had become so overused that it was a chief cause of national agony.

□   □   □

One agronomist, living and working during the 1930s in Ohio, at the eastern edge of "the nation's great endowment," decided that the agony of soil erosion and agricultural depression was not fore-ordained, could have been avoided, and even now was remediable, if only farmers would give up the moldboard plow. The agrono-mist's name was Edward H. Faulkner, and there ought to be a great, green statue of him in front of the Department of Agriculture's building in Washington. Hardly likely. Faulkner had to resign from the department to pursue his theories, which he published in 1943 in a small book called *Plowman's Folly*. Though a best-seller, the book, and its author along with it, are all but forgotten by a new generation of government and academic agricultural experts, many of them of the hidebound sort that Faulkner would probably still be doing battle with were he alive today.[3]

In 1943, the United States was in the middle of a war that nobody was yet entirely sure it would win, or at least win hands down. The Dust Bowl droughts were gone, but not forgotten; there was tre-mendous pressure placed on agriculture to produce more and more for the boys at arms and for starving allies, even as farm manpower was reduced by enlistments and the draft (though there were agri-cultural deferments). Food rationing was in force, and forty million victory gardens had been established on the home front by patriotic American families to supplement the diminished amount of fresh produce available at the grocery store. Interest in the science and art of agriculture was, therefore, at its highest point ever, even among those who hadn't given it a second thought before.

This was the context in which Edward Faulkner, with a kind of cool daring, made these assertions in the opening paragraphs of *Plowman's Folly*, which are quoted at some length so that Faulkner's central argument can be presented in his own words:

> Briefly, this book sets out to show that the moldboard plow, which is in use on farms throughout the civilized world, is the least satis-factory implement for the preparation of land for the production of crops. . . . The truth is that no one has ever advanced a scientific reason for plowing. . . . The entire body of "reasoning" about the management of the soil has been based upon the axiomatic assump-tion of the correctness of plowing. But plowing is not correct. Hence, the main premise being untenable, we may rightly question the valid-

---

[3]Faulkner was born in 1886 in Whitley County, Kentucky, and died in 1964 in Elyria, Ohio. The date of his birth, should any public-spirited citizen concerned with the improvement of American agriculture wish to declare a "Faulkner Day," is July 31.

ity of every popularly accepted theory concerned with the production
of any crop, when the land has been plowed in preparation for its
growth.

. . . In brief, if a way had been found to mix into the surface of the
soil everything that the farmer now plows under; if the implements
used in planting and cultivating the crop had been designed to operate in the trashy surface that would have resulted from mixing rough
straw, leaves, stalks, stubble, weeds, and briars into the surface—
crop production would have been so automatic, so spontaneous that
there might not have developed what we now know as agricultural
science. . . . That we would also have missed all of the erosion, the
sour soils, the mounting floods, the lowering water table, the vanish-
ing wild life, the compact and impervious soil surfaces is scarcely an
incidental consideration. . . .

. . . We have, by plowing, made it impossible for our farm crops
to do their best. Obviously, it seems that the time has arrived for us
to look into our methods of soil managment, with a view to copying
the surface situation we find in forest and field where the plow has
not disturbed the soil. No crime is involved in plagiarizing nature's
ways. Discovering the underlying principles involved and carrying
them over for use on cultivated land violates no patents or copyrights.
In fact, all that is necessary to do—if we want a better agriculture—
is to recharge the soil surface with materials that will rot. Natural
processes will do the rest. The plant kingdom is organized to clothe
the earth with greenery, and, wherever man does not disturb it, the
entire surface usually is well covered. The task of this book is to show
that our soil problems have been to a considerable extent psycholog-
ical; that, except for our sabotage of nature's design for growth, there
is no soil problem.[4]

This remarkable book is, in every important respect, the theoretical cornerstone of conservation tillage, though only now is its
author beginning to get some recognition for his extraordinary contribution. It might never have happened, for the book very nearly
did not get published.

One day, at Malabar Farm, the thousand-acre Ohio establishment owned by Louis Bromfield, millionaire novelist, experimental farmer, and soil conservationist, a "smallish, graying man, with
very bright blue eyes" presented himself to the well-known author,
who had been visited already by a procession of cranks proferring
endless horticultural panaceas. "He said his name was Faulkner,"
Bromfield continues (in the first of his nonfiction farm books, *Pleasant Valley*), "and that he wanted to talk about a new theory of cul-

---

[4]From *Plowman's Folly*, by Edward Faulkner. Copyright 1943 by the University
of Oklahoma Press.

It was to this imposing Ohio farmhouse, Louis Bromfield's Malabar Farm, that an unknown agronomist with a "crazy idea" brought the manuscript for a book called *Plowman's Folly*. The idea was to do away with the moldboard plow altogether and leave crop residues on or near the surface to improve the tilth of the soil instead of turning it under, as was almost universally recommended by agricultural experts. The book, finally published in 1943 with the help of Bromfield and others, became a best-seller. And its author, Edward Faulkner, is now credited as being the "father" of conservation tillage.

tivation that did away altogether with the conventional, long-accepted moldboard plow." Bromfield listened to Faulkner's explanation for a while, mumbled something perfunctory, and bade his visitor goodbye, hoping he might never see him again. Undaunted, Faulkner returned several times, insisting that Bromfield pay attention. Finally, he did. "I had never thought of the evil the moldboard plow might do," Bromfield confessed, "until I listened to Ed Faulkner."

In due course, Faulkner presented Bromfield with a long manuscript for a book and asked for help. The proprietor of Malabar had his own literary projects to attend to, necessary for subsidizing the farm, but he sent Faulkner to, among others, Paul Sears, author of the 1930s classic *Deserts on the March*. Sears and others worked on the manuscript and, after offering it to five publishers who turned it down, finally persuaded Savoie Lottinville of the University of Oklahoma Press to take it on. What happened next is a fascinating footnote in publishing history. Oklahoma published a much-shortened version, owing to the wartime paper shortage as well as to literary considerations for a book that was, after all, quite technical and unlikely to get much attention from either the general public or agricultural authorities. As it turned out, Oklahoma went through eight printings in scarcely more than a year. Then, out of paper, they turned to Grossett & Dunlap, a major New York trade publisher, who printed 250,000 copies in 1944, an astonishing number for a book that ordinarily might be expected to sell no more than 5,000 copies. But the book was a hit from the very beginning. Articles about it were published in magazines from the *New Yorker* to the *Saturday Evening Post*. Bromfield himself wrote some of them and was accosted (if that is the right word) by two Hollywood actresses in a Chicago hotel, who asked, "What is all this business about *Plowman's Folly?*"

The business was a serious one, for the moldboard plow had done, and continued to do, great damage to hundreds of millions of acres of American agricultural land. What Faulkner was arguing for was not the total absence of tillage, but a tillage done with means other than the moldboard plow, which, though it had its place in the clearing of land, was inappropriate as a tillage tool after that. Traditionally, the act of turning the soil via the moldboard plow was considered simply to be a field-scale version of preparing garden soil by means of a spade. However, the preparation of garden soil is an *intensive* form of cultivation wherein most essential natural processes are replaced by the gardener in enriching and aerating his or her plot. Drawing a moldboard plow through a field merely interrupts the natural processes of undisturbed soil rather than replacing them with the gardener's artifice. What is worse, the plow not only destroys the natural functions of this soil but actively inhibits the future capacity of the soil to heal itself.

In its natural state, soil is, as Faulkner reminded his readers, a quite delicate piece of work. Microscopic organisms labor ceaselessly on bits of plant material such as stalks, roots, and leaves, reducing them to nutriment and giving the soil its dark, rich "tilth"; small animals, like earthworms, dig down beneath the frost line in winter, opening tiny channels through the soil to deliver water to the surface, to make trace elements available, and to provide essential pathways to the subsoil for new rootlets; the surface blanket of organic matter protects this feverish activity by providing shade and holding in moisture, though allowing the surface of the soil to breathe—and all this is violently disrupted by a single pass of a moldboard plow. The only human-scale analogy that comes to mind is the awesome destruction that obtains from the geological shift of great tectonic plates, producing an equivalently violent wrenching of the earth that ruptures highways, disconnects gas and water mains, topples bridges, and buries neighborhoods in rubble. Just as the whole infrastructure of the human community can be disrupted by such gigantic seismic events (as in Mexico City in 1985), so the smaller-scale, but no less important, infrastructure of the *soil* community is catastrophically damaged and deranged by the tearing, slicing action of the plow, which not only cuts through the soil but lifts it, turns it, and dumps it upside down.

Abandoned farms were a common sight in the 1930s. It was due, said Faulkner, to destroying, by means of the moldboard plow, the natural capacity of the soil to maintain its "self-sufficiency." Faulkner's experiments showed that land was "farmed out" not because of lack of nutrients in the soil but because its natural processes were interrupted by plowing, which led to compaction (hardpan) and erosion.

As Faulkner describes it, the cutting and turning action of the plow means that "soil simply takes time out from the business of growing." The plow, he says, destroys the natural particle-to-particle capillary action that delivers moisture from the subsoil to the surface for plant roots to use. "Every clod," says Faulkner, referring to the effect of plowing when the ground is wet, "is so much soil mustered out of service for the season."

Moreover, there is the serious matter of hardpan or plow pan, a form of subsurface "clodding" that is a by-product of continuous moldboard plowing and that can become a permanent feature of the soil's "profile." Plow pan is formed just below the level of the plowshare, and its effect is to deny plants the subsurface water they need because, using Faulkner's term, the natural "wicking action" of the soil is sealed off. The "seal" works the other way around, too, by inhibiting the drainage of rainwater into the subsoil, since the plow pan creates a nearly impermeable barrier that causes the surface water either to stand on a field if it is flat, literally drowning a crop, or to run off if the field slopes, carrying topsoil with it. The field, in the extreme case, becomes a thin layer of pseudo-soil, as if it were a layer of dirt spread on a concrete slab. Nothing works. The plow pan inhibits the ability of the plant roots to go deep enough to find water and essential nutrients. The crops get smaller each year, becoming more susceptible to disease, plagues of insects, and weeds. At some point, the farmer of old would shrug his shoulders and move on, claiming that the land was "farmed out," when in fact, as Faulkner would argue, the land had not properly been farmed at all.

□   □   □

Faulkner not only showed how the moldboard plow destroyed the soil's natural ability to grow crops when left more or less to its own devices, but he also set forth a system of farming that came to be called the "Faulkner method." Some subsidiary aspects of the method had less universality than Faulkner originally claimed for them (in a 1947 book called *A Second Look,* he modified some of his suggestions), but the subsidiary principles should not be dismissed entirely, even though nearly half a century later, they still sound revolutionary.

Faulkner advocated using a disk harrow (though he admitted it was not a perfect implement) to mix residues, such as corn stalks after harvesting, into the top layer of the soil—no more than two or three inches deep. In this way, the field became in effect a giant compost heap with a kind of "instant" topsoil created or recreated on poor land. By "mixing it in from the top," the basic structure of the soil was not damaged, even though its "organic matter profile" (OMP) was improved.

One of the reasons for lower yields when organic matter is buried rather than mixed is that the buried material simply putrefies, creating heat, which tends to destroy the beneficial organisms that break down the organic matter so that it can become integral with the soil itself. Thus, turning under the residue, as would a moldboard plow, not only is ineffective, it works *against* an improved OMP.

The benefits derived from Faulkner's tillage methods were three. First, much less effort was needed in the process of cultivation. As Kentucky no-till pioneer Harry Young puts it, "Moving 1,000 tons of soil per acre while plowing only 8 inches deep, moldboarding takes plenty of time, machinery, and costly energy."[5] The second benefit was the development of "self-sufficient soils" for the farmer and those who would use the land after him. And third was consistently improved yields—because more nutrients could reach the plants; because there would be a steady supply of moisture (rather than alternations of flooded and dried-out soil); because stronger

---

[5] Young, "the father of no-till," planted his first no-till crop in 1961. He is considered to be the first commercial farmer to use this method. See the citation in List of Principal Sources for his no-till handbook.

roots could go deep to get trace minerals from subsoils; and because of renewed availability of nitrogen, which Faulkner said was a natural by-product of his method, among other reasons.

On the matter of fertilization, Faulkner believed (and it got him into much trouble with the agricultural establishment) that in humid areas with loams such as those found in Ohio, "there is absolutely no need for commercial fertilizer; nature can make available annually enough new plant food to grow crops several times as large as we produce now." On the matter of insect pests, he believed (more trouble) that his method so strengthened the plants that they were materially less susceptible to damage and that the plant's juices were so rich in minerals and so lacking in sugar (a sign of plant weakness) as to be unpalatable to insects. "It becomes possible," he concluded, "to improve the human food supply by the very method that will starve the insect."

Faulkner also believed, and here we come to the heart of the matter, that his method could lead to "weedless farming":

> Seed the land to a green manure crop; rye in the fall, or a suitable summer crop in the spring. Let the green manure crop grow until it has reached the proper height to be worked into the soil with the available equipment. If weeds growing in the green manure crop begin to bloom, it is important that the crop be put into the land immediately. However, few weeds mature quickly enough to rush the incorporation of the green manure. Under almost all ordinary farm conditions in the humid section of the country, it will be possible to grow a winter crop and a summer crop, put each into the soil with its accompanying immature weeds, and in a short time bring about fertility of the soil and at the same time help rid the land of weeds that create the necessity for cultivating farm crops.

The point is that by not turning up the soil, as with a moldboard plow, the weed seeds of other seasons are not exposed and therefore do not germinate. By putting the land into what Faulkner describes as a rapid succession of winter and summer green-manure crops, which are disked before new weed seeds are formed, eventually all the old weed seeds near the surface will sprout; no new weed seeds will be introduced; the soil will be enriched; and eventually—within two to five years, said Faulkner—the field will be weed free. If, as Faulkner put it, "weeds can be so controlled that the farmer's crops are not forced to compete with them for the plant food in the soil, then it goes without saying that no cultivation should be undertaken."

□   □   □

To report that these ideas were greeted by skepticism in the agricultural scientific community of the time is a massive understate-

Stubble-mulch tillage originated in the Great Plains during the 1930s and 1940s and is considered to be a forerunner of conservation tillage, along with Faulkner's "plowless farming" techniques described in *Plowman's Folly*. This photograph shows a modern version of stubble-mulch tillage, using a chisel plow with straight shanks going deep to break up compacted subsoil.

ment. Had his book not been a best-seller, it would have been ignored. But it was a best-seller, and so Faulkner was branded a "nut," a "crank," and a "fanatic" whose basic theory of natural soil dynamics was dismissed because subsidiary elements, such as the self-generation of nitrogen and the plant-juice business, could not sustain (perhaps for good reason) scientific scrutiny. In *A Second Look*, Faulkner tells of attending a major scientific meeting on soils. After listening to all the speakers, he bitterly concluded: "It seems certain that none of these men believes the soil can be self-sufficient. Each speaker assumed just the opposite in fact. To this extent, therefore, the thesis of *Plowman's Folly* has no standing with this group." The scientific community had unfortunately confused Faulkner's basic ecological theory—among the most percipient ever advanced in agricultural science—with his assertions of potential subsidiary attributes.

The big issue was, and still is, weeds. While "stubble-mulch tillage" did begin to develop during the 1930s and 1940s in dryland farming areas where weeds were less of a problem, Faulkner's "plowless farming" was thought impractical in humid areas simply because there was no obvious way to control weeds except by plowing them down in the fall and again in the spring with a moldboard and then cultivating betweentimes. In some areas, however, chisel plows were used to prepare the soil for planting instead of moldboarding and harrowing. Like Faulkner's disking, chisel plows did not actually turn the soil, bringing weed seeds to the surface, but they could go deep, incorporating manures, and in some versions called "subsoilers," even breaking up the hardpan. This was one of Louis Bromfield's adaptations of Faulkner's theories, along with the use of a large rototilling machine.

While such techniques could ameliorate the weed problems, they could never really eliminate them, or at least deal with them as decisively as in a "clean-tilled" field, and soon, except among a handful of followers (like Bromfield), Faulkner was forgotten. His star waned almost as quickly as it had risen. Instead, it became the agricultural chemist's finest hour in the history of conservation tillage. According to Robert Rice, author of a conservation-tillage textbook subsidized by Chevron Corporation, makers of Ortho agricultural chemicals, the "revolutionary alternative to tillage" really took shape in the laboratory in the 1940s, not in Faulkner's test plots. "The technology that made this possible," Rice says, "was a by-product of World War II research—the discovery of 2,4-D and related organic phenoxy herbicides." However, while 2,4-D was effective (in fact, it was an active ingredient, along with 2,4,5-T, of Agent Orange, the controversial defoliant used in the Vietnam War) and widely used on broadleaf weeds, it did nothing to kill grasses— especially the pesky quack grass and foxtail.

As Purdue University soils scientist William C. Moldenhauer recounts this period in the history of conservation tillage, "When herbicides came in and you could control the weeds without having to till the hell out of the ground, we began to have more confidence in our ability to leave some residue on the fields." The next improvement in herbicide effectiveness was atrazine, which could control a broader spectrum of weeds and accordingly gave another boost to conservation tillage. But the really big breakthrough, according to Moldenhauer, was not 2,4-D or atrazine, but paraquat, a chemical first imported by Chevron from Great Britain's giant chemical manufacturer, ICI, in the early 1960s. Paraquat is a "burn-down" herbicide that does its job quickly, killing every plant it touches but becoming botanically inactive when it hits the soil. Planting on a "burned-down" field can theoretically be done the same day as the spraying. It really worked, and it still does. Says Donald R. Griffith, a Purdue extension specialist and like Moldenhauer, an early conservation-tillage researcher, "There really wasn't much doing in the way of conservation tillage until paraquat came along about twenty years ago."

□ □ □

To an outsider—a nonfarmer—it may seem a mystery that Edward Faulkner is accorded so little scientific recognition for his contribution to the practice of agriculture in general and to conservation tillage in particular. But that omission may be in the process of correcting itself, despite the long shadow cast by the role of herbicides in bringing the "revolution" about.

Herbicides were, and still are, essential in moving the concept of conservation tillage from a set of theories into actual practice in American agriculture. But in the long run, seeing conservation tillage simply as the substitution of herbicides for tillage is a too-simple view of a complex of technical factors that come into play. For one thing, a whole new approach to tillage machinery design has been developed, which lagged behind the development of herbicides. Moreover, new rotations, new ways of planting, new fertilizers and fertilizer-application techniques, even new crops must be part of a conservation-tillage "system." When all these factors are in place, then the theoretical debt that conservation tillage owes to Faulkner becomes plainer, and some scientists are beginning to recognize this. On the matter of soil capillaries, for example—the "wicking action," as Faulkner described it—soils expert William Moldenhauer perhaps speaks for more scientists than himself when he says: "We're back to that now. We didn't pay attention, and we should have."

In fact, William Moldenhauer, who retired in 1985 as head of the National Soil Erosion Laboratory and is generally recognized as the dean of conservation-tillage research scientists, says that he was inspired by Faulkner to go into agricultural science in the first place. But then, says Moldenhauer, he became disenchanted: "All through my bachelor's, master's, and Ph.D. studies, they said he was a nut." But now he is reenchanted: "I went back and read his book a month ago [in 1986], and have come to the conclusion that if Faulkner had all the information we have today, he could have been right on. Faulkner told us what we ought to do. But we were not able to do it then. Now we can."

And so the mad theories, which found their origin in the work of Edward H. Faulkner, are being worked out at field scale by conservation tillers all over the country. The leading newsletter in the field, *No-Till Farmer*, carries the Faulkner quotation "No one has ever advanced a scientific reason for plowing" in every issue.

□ □ □

Conservation tillage came into its own only after World War II and the development of sophisticated herbicides. And yet, essential as herbicides are to some forms of conservation tillage, as this photograph suggests, the new way of farming is far more complex than simply substituting herbicides for tillage.

A final point. There may be another reason it has taken so long for Faulkner to be recognized: his message was conveyed in largely negative terms, as the very title of his book demonstrates: the plowman's *folly*. Moreover, his rhetoric was uniformly clumsy: "working it in from the top" was about the best he could do to describe his process of residue management. A writer for *Country Gentleman* magazine did a little better, but not much, in the title of a 1939 article about Faulkner—"Right Side Up Farming." But farmers are conventional souls, easily affected by ridicule. Should the fellows at the co-op ask a Faulkner adherent what in the world he thought he was doing not fitting up his fields like everyone else, how possibly could he reply that he was engaging in "right-side-up farming" by "working it in from the top?" It is no wonder that during the 1940s, 1950s, and 1960s, the nomenclature for not using the moldboard plow became so complicated that the technical differences, rather than similarities, seemed more important: "plow-plant," "no-till," "mulch tillage," "till planting" and others were defined and redefined (and still are) in terms of comparison with one another rather than in terms of their *effect*.

But in one of history's happy circumstances, a solution to this problem was provided that turned Faulkner's folly around, putting his theories, and the improvements others made upon them, in positive rather than negative or purely technical terms. According to Vincent J. Price of the Soil Conservation Service: "In 1967 in Illinois, the Soil Conservation Service, the Cooperative Extension Service, and the Agricultural Stabilization and Conservation Service, through the Illinois Conservation Practices Committee, launched a program to promote minimum tillage. First they settled on a name for the practice—*conservation tillage*." Whether all those agricultural bureaucrats knew it or not, they did everyone a great service with this most resonant and affirmative term, conservation tillage. Within it all the variations on not using the moldboard plow and keeping organic residues on or near the surface of the field to curb erosion, conserve soil moisture, and improve the self-sufficiency of the soil can be collected under a single rubric. Not everyone is yet willing to accept conservation tillage as the flagship generic term, but its time is coming, just as sure as the practice will eventually replace the plow, despite Virgil, Jethro Tull, Thomas Jefferson, John Deere, and even Ed Faulkner himself.

4

Old Yellow

I was born on the prairie and the milk of its wheat,
the red of its clover,
The eyes of its women, gave me a song and a slogan.
—CARL SANDBURG

From the time of the stubble-mulch tillage experiments on the Great Plains and the work of Edward Faulkner and his followers in the Midwest right up to the present day, the development of conservation tillage has been based on the land, in real soil on real farms. The new tillage procedures were conceived of for practical reasons, and new implements were invented—by farmers, more often than not—to give the new ideas a trial. And what worked, worked. Later, the agricultural scientists would arrive to try to figure out how come and set up their test plots and write their papers. Manufacturers would mass-produce the machinery. And soon the techniques would evolve enough to be taught in the agricultural schools. The danger with this is that the concepts tend to become abstracted, disconnected from the land whence they arose.

Perhaps the most curious aspect of moldboard-plow agriculture—looking backward with the clarity of hindsight—is that it treats all land the same. Or perhaps better put, it treats the land almost as though it did not exist. No matter that the rain falls hard upon it or not at all. No matter that its soil is made up of layers of glacial till, or of aeolian loess, or of a former seabed. No matter if the land's fields are flat or tilted; if a plow could cut through them, most likely a plow did so, without regard to the vast regional dif-

Wheat is a principal commodity crop in the "dryland" farming areas west of the 100th meridian. In the West, where the economic penalties for erosion are severe, conservation tillage can make an important contribution.

ferences in the physical geography of a continent-sized agricultural land base.

In contrast, conservation tillage, by its very nature, *begins* with the land. To grow a good crop, the techniques of conservation tillage must be mated to the particularities of the ground it is to be practiced upon: the moisture, density, organic content, and the host of other factors that make up the soil's physical profile, as well as to peculiarities of what the land will grow and what it won't. This is because conservation tillage must work *with* soil, letting the soil handle the task of growing plants as it was meant to.

Accordingly, the differences in land are the reason why there are different versions of conservation tillage, different terminology to describe them, and most particularly, different machinery for their implementation. The variations in conservation tillage are not, as partisans of one or another of its versions would have you believe, entirely a matter of happy chance or a greater endowment of wit or enterprise on the part of its practitioners. Instead, the various types of conservation tillage derive from basic biophysical differences in the regions in which commercial, commodity-crop agriculture is carried out. And so, fully to understand conservation tillage, one has to view it in situ—in the agricultural landscapes that have given rise to particular tillage techniques.

We have already discussed one of them. The Maryland fields of Todd Greenstone, who uses a version of no-till, are generally rep-

resentative of the light piedmont soils of the northeastern and middle Atlantic states. What may be called "Eastern no-till" requires a
planter that is only slightly modified from the one used for planting
a clean-tilled field under a conventional tillage regime. Essentially,
herbicides are substituted for fall and spring plowing and for interim
cultivation. Winter frosts guard against excessive compaction, the
abundant rains are absorbed into the subsoil, and the microorganisms quickly break down the residues into usable materials to add
tilth to the soil. But elsewhere, the biophysical environment is different, and accordingly, different systems and quite different
machinery have emerged. With regard to no-till, the western techniques developed for wheat production in hard-soil dryland farming areas are altogether unlike the methods and machinery required
for Todd Greenstone's gently rolling Maryland countryside.

□   □   □

Dryland farming[1] is the primary agriculture in virtually all parts
of the West. While some 46 million acres of western croplands are
irrigated, the bulk of its farmland is not. In fact, there are 156
million acres in dryland farming in the western United States, a
statistic that surprises many people. Most think of western agriculture as being mainly rangeland, with cropland confined to those
areas capable of being irrigated. The fact is that better than a third
of all U.S. land used for crops is in the "dryland" category—unir-
rigated land west of the 100th meridian.

If you draw a vertical line down the face of the United States
from, roughly, the middle of North Dakota through, roughly, the
middle of Texas (with the Panhandle to the left of the line), you have
described this famous imaginary boundary—the 100th meridian—
west of which is the real West and the dryland farms. Except for
the northern part of the coast and the Sierra Nevada–Cascades,
this is arid country, the American steppe, where a maximum of 20
inches of rain falls per year, and usually much less. By contrast,
most of the midwestern, eastern, and southern United States gets
between 40 and 100 inches of rain per year.

Corn doesn't grow west of the 100th meridian, nor do soybeans,
except on the irrigated lands. This is small-grain country—wheat,
barley, sorghum, and millet. Lentils, dry peas, safflowers, and sun-
flowers also grow here, alternating with the grains—from the pan-
handles of Oklahoma and Texas up through western Kansas,

---

[1] Dryland farming is defined by the U.S. Department of Agriculture as "a system
of producing crops in semiarid regions—usually with less than 20 inches of annual
rainfall—without the use of irrigation. Frequently, in alternate years part of the
land will lie fallow to conserve moisture."

In contrast to conventional tillage, in which land is simply plowed, the various forms of conservation tillage begin with the local particularities of crops, soils, and environmental factors such as rainfall. Here a scientist measures soil moisture with a neutron probe that accurately measures moisture to a depth of over six feet. In dry areas of the country, deep soil moisture can spell the difference between success and failure.

Nebraska, and the Dakotas and into the prairie provinces of Canada. West of the Rockies, dryland farming picks up again in the basin-and-range country running to the Sierra-Cascades, in parts of Nevada, northern California, Utah, Idaho, Oregon, and Washington. There's even some dryland farming in southern California, though mostly that state is irrigated where it is not desert or mountain.

In all these places, erosion is of immediate economic concern to everyone. This is no textbook subject for the dryland farmer. Erosion directly decreases the ability of the soil to retain scarce moisture in arid and semiarid regions, and where the moisture is lacking, the crops don't grow well. Moreover, the effect of erosion on limiting the ability of the soil to retain moisture is basically irreversible. And this difficulty is especially pronounced in the steep hills of the Palouse country, an astonishingly productive wheat-growing region located in the southeastern corner of Washington state and spilling over into Idaho.

□   □   □

The Palouse is perhaps the most starkly beautiful agricultural landscape in North America. Geologically, it is in a category of one, sui generis—a fertile, deeply folded, treeless prairie consisting primarily of loess soil piled more than a hundred feet deep in places. During the Pleistocene age, the great sheets of ice moving across Canada and into the northern tier of the United States did not reach the Palouse, but the weather changes that followed whipped up volcanic ash from the forebears of Mount St. Helens and, swirling the gray, siltlike material in clouds of great tonnages, deposited it, along with sediments of glacial origin, in steep, random, dunelike formations covering some two million acres—an area about half the size of New Jersey.

The good news about the soil of the Palouse, the wind-borne loess (from a German word meaning "loose"), is that it is enormously fertile. It is friable, deep, almost like potting soil in the best places. Today, the region ships most of its soft, white wheat to the Orient for noodles and produces additional white and red wheat for sale elsewhere. The bad news about this soil is that, like the sand dunes it resembles, it is prone, when opened by the plow, to severe erosion from wind and water. Loess is not confined to the Palouse, of course. Loess deposits constitute the deep, rich bottomlands along the Missouri and Mississippi rivers—areas where erosion rates are also extremely high. But in the Palouse, the erosion, though confined to a relatively small area, is the most dramatic of any place in America. According to Frederic R. Steiner of Washington State University (located in the heart of the Palouse, at Pullman), some of the steeper slopes (some with as much as a 50 percent grade) suffer erosion rates between one hundred and two hundred tons per acre per year. The U.S. Department of Agriculture's "allowable" rate—the amount of topsoil loss that does not permanently damage cropland—is five tons per acre per year.

The Palouse has not been farmed for long—scarcely more than one hundred years. When first settled in the mid-1800s, the land was considered too dry for farming, so it was used as cattle and sheep range.[2] Then, beginning in the late 1870s, farmers were attracted to the region by the opening of the Northern Pacific Railroad, and they soon discovered how fertile the soil beneath the native bunchgrass and snowberry really was. They began "dry farming" the steep landscape, plowing deep to catch the rain and cultivating frequently betweentimes to keep the surface soft and

---

[2]The name Palouse may come from the French *pelouse,* a grassy expanse, or from the Nez Perce Indian word for the area, *Palloats.*

The Palouse country of eastern Washington state and western Idaho is a unique and starkly beautiful prairie landscape consisting of dunelike formations of fertile aeolean loess. It is also, as the photograph opposite shows, one of the most erosion-prone agricultural regions in the United States.

accessible to moisture. The newly built railroad brought even more farmers to the Palouse, and before long a substantial part of the region was converted from rangelands to farms—mixed farms at first; then, specialization in small grains began. Now winter wheat is the principal cash crop, with spring wheat, barley, and peas and lentils following.

The result of this kind of agriculture, says Steiner—growing wheat by "plowing for rain" on the tilted fields of the Palouse for a hundred years—is that 100 percent of the original topsoil has been lost from one cropland acre out of ten, and somewhere between 50 and 75 percent of the topsoil has been lost on a majority (six acres out of ten) of the rest of it. The Palouse, therefore, is one of conservation tillage's toughest challenges. If it can stem erosion here—and consequently increase yields and income for the farmer—it can be effective almost anywhere on that crucial third of American cropland, the dryland farms of the West.

To determine the nature of the connection between erosion and yield in dryland wheat-growing areas, two U.S. Department of Agriculture researchers, T. W. Massee and H. O. Waggoner, conducted two series of tests—one on "artificial" plots to demonstrate the effect of erosion on a range of western dryland soils and the other on actual farm plots in northern Idaho, not far from the Palouse. In the "artificial" tests, Massee and Waggoner compared an ideal soil (created by adding six inches of topsoil, thus replicating, perhaps, what early farmers originally found in places like the Palouse) with existing "untreated" soils and those with six or twelve inches of topsoil removed. The results were "large and significant

yield differences," in the view of the investigators. Compared with the plots with the ideal soil, wheat yields declined from 45 bushels per acre when no fertilizer was added to 27 bushels per acre on the untreated soil and 14 and 11 bushels per acre, respectively, on soil with six and twelve inches of topsoil removed. Interestingly, while fertilizer could offset (but not compensate for) reduced topsoil in the artificial plots, this was not necessarily the case on the actual farm-field tests, where fertilization could not overcome the droughtlike conditions that obtain in eroded soils when water is less able to infiltrate the soil and therefore simply runs off the steep slopes of the fields. Wheat yields on the eroded fields were half what they were on uneroded fields.

The key factor is moisture: without it, plants grow poorly, and fertilizer doesn't help much.[3] There is less soil moisture on eroded land than on uneroded land. A kind of artificial drought is set up in the soil because the moisture cannot infiltrate the less-permeable surfaces of eroded soils. Moreover, land where topsoil has been eroded away is more subject to erosion than land where topsoil is more or less intact. Thus a vicious circle is set up in dryland farming in which irrigation cannot be used to compensate for lack of moisture: the more erosion, the less moisture; the less moisture, the less plant growth; the less plant growth, the more erosion. In its extreme form, this is the round dance that leads to desertification.

---

[3] On the test plot replicating deep topsoil, the researchers found that the natural nitrogen did not need to be augmented by commercial fertilizer, which, when applied, did not significantly increase yields.

And that is where Mort Swanson and Old Yellow enter the story.

Swanson is a Palouse native, sixtyish though scarcely looking it, and has been farming land near Colfax, Washington, for forty years. He is a compact and wiry man with bright, button eyes, and he has a precise way of speaking. His office is a partitioned-off corner of a huge, hangar-sized shop with over five thousand square feet of heated space, complete with parts bins, welding equipment, overhead cranes, work benches of every description, and a sizable open workspace in the center.

Though plain and unpretentious, Swanson's office does not look like a farmer's, for it is dominated by a large, commercial-sized engineering drafting table rather than untidy stacks of old fertilizer brochures on broken chairs, or piles of muddy field clothes thrown in a corner, as might be the case with most farmers, to whom office work is unpleasant and to be avoided. Not this farmer. On many a late night, bent over this very drafting table, Mort Swanson created the detailed plans for a reliable and rugged machine that would provide a way to no-till not just the soft soil in the swales, but the characteristic hard-dirt steeps of the Palouse. Swanson is a believer. "No-till," he says, "is the best thing that happened to conservation since the soil conservation movement started."

The product of Swanson's effort was "Old Yellow," a huge, no-till seed drill built right in the shop and first used for the 1974 planting season. Swanson realized that adapting existing seed drills for no-till on the hard, unplowed surfaces of the Palouse hills—especially on the eroded knolls—simply would not work. "We built Old Yellow

This Yielder no-till drill is much the same as Morton Swanson's original design, nicknamed "Old Yellow." It weighs in at twenty-five tons fully loaded and requires the largest-size tractor to pull it up the steep hillsides of the Palouse.

from scratch," says Swanson. "We didn't even think about modifying existing equipment." What he built was a twenty-foot-wide planter with huge seed-carrying bins and fertilizer tanks on top and, underneath, the most astonishing assortment of gadgets—coulters, disks, and packer wheels to cut, open, or close the soil; tubes to deliver several kinds of fertilizer; hydraulic lifters; springs; valves; sensing devices; sprockets; chains; piping; hitches; and steel rods, plates, brackets, and beams to hold it all together—since Rube Goldberg's very worst nightmare. But it works.

Swanson says that it all came about because in the fall of 1973, there was a field he planned to seed with winter wheat, but he was late, and instead of plowing down first as on other fields, he had simply planted into the stubble with his regular seed drill, albeit with some difficulty. Even so, "That was the best crop of wheat I had raised up to that time," he says. There and then, he decided to build a seed drill that could effectively reproduce those results, not with difficulty on a single field, but easily and reliably for a whole no-till operation. In Swanson's case, this amounts to twelve hundred acres.

Old Yellow and its descendants weigh up to fifteen tons—twenty-five tons when loaded with seed, herbicides, and fertilizer. "There are times you need every bit of that weight to get penetration," says Swanson, referring to the hard, dry surface of Palouse soil as well as the difficulty in cutting through the tough, dense wheat stubble from a previous crop. Moreover, he says, the machine has to be heavy to withstand the enormous twisting pressures exerted on the disk openers and the chassis of the machine on the steeper hills. Palouse farmers do not go around anything, but plant up hill and down dale despite seemingly impossible grades. "When you get on a slope of 50 percent," says Swanson, "it really puts a strain on everything." The tractor required to pull such a machine is equivalently impressive—Swanson's is a four-hundred-horsepower Steiger diesel, driving eight heavy-ribbed tires, each as high as a man is tall.

What Swanson's no-till drill does is actually fairly simple. It cuts slots for seeds through the stubble and a deeper slot for fertilizer, and it can place additional fertilizers or herbicides, if required, and then cover the whole business up, leaving a stand of last season's stubble relatively intact, yet perfectly planted with a new crop. "In no-till," says Swanson, "the roots from the previous crop die right in place and leave all those holes where the roots are so that water has a way to get into the soil." Thus, when the rains come, moisture can get down to the roots of the new crop because the roots of the old crop were not destroyed. In time, the surface residue breaks

Close-up view of the business end of an Old Yellow–type drill, showing coulters, lifter valves, tubes, and packer wheels affixed to "tool bars."

down—"faster on the surface than it does when it's buried," says Swanson—and adds permanent organic matter to the soil, increasing its inherent ability to hold moisture.

The basic structural concept of Old Yellow was the use of a series of "floating tool bars" of two-by-six-inch steel on which to affix the various disks, delivery tubes, and packer wheels in a variety of arrays to meet particular planting needs. One tool-bar array option involved an implement to permit fertilizer to be placed beneath the surface of the soil to eliminate wasteful broadcast methods of fertilizer application. On this topic, Swanson had long discussions with the Colfax county agricultural agent, who proposed "shanking in" the fertilizer by affixing a curved steel implement on one of the front tool bars. The shank would act as a kind of very narrow plow that could cut a deep slice into the ground so that fertilizer could be placed below the seed, where presumably it would do the most good. Swanson was leery, however, of the loose seedbed that this technique would produce—leading to compaction by his heavy equipment, disrupting the soil structure, and uprooting the stubble that served to inhibit erosion on the steep slopes. His alternative idea was to make a V-shaped opening (with offset coulters) for the fertilizer to the side of the seed row, figuring that the lateral roots of the young wheat plants could reach the fertilizer about as well as with the shanking method but without so much disturbance of the soil.

In a test of this method, however, one of Swanson's employees, ordered to rearrange the devices on the tool bar, misaligned the sets of disks that were to make the shallow cuts in the soil for seed placement and the somewhat deeper "V" for the fertilizer. The result was that instead of rows of wheat equidistant from the fertilizer trench and from one another, the seeds were placed in pairs of rows only five inches apart with fifteen inches between the pairs. What was worse, the fertilizer opening was crammed between the seed rows so that only the lateral roots on one side of the plant could receive the nutrient. The mistake had been made on a big field, so all Swanson could do at that point was to wait and see what the damage would be to his yield.

Mistakes like this should be made more often. The "paired-row" field, with the deep-banded fertilizer *between* the narrow rows, out-produced all of Swanson's other fields. The reasons were several. First, the fertilizer was so localized that it got the young seedlings off to a faster start, despite its eccentric location, resulting in more overall vigor. Second, the paired rows, being so close together, provided effective crop competition against weeds, crowding them out of the immediate growing area. Third, since the space between the row pairs was unfertilized, weeds in this area had less vigor than would be the case if the fertilizer were evenly distributed. Fourth, a greater amount of stubble from the previous crop was left undisturbed, producing advantages in overall tilth, reduction of compaction, and control of erosion.

While there had been general recognition that it was better to "feed the wheat and starve the weeds," as writer Glenn Lorang put it in a *Farm Journal* article on fertilizer placement, only the very solid and precise kind of no-till drill that Swanson had invented could produce what is now generally known as paired-row planting. According to Robert I. Papendick, a USDA soil scientist from the Washington State University at nearby Pullman: "We have always known banding would increase yields. But until now, we haven't had the equipment to get through the really heavy wheat stubble and put the fertilizer where it needs to go." In controlled tests at Washington State, Papendick found that the paired-row technique could increase yields of winter wheat by 13 percent and of winter barley by 9 percent.

Moreover, since the row pairs are separated from one other, herbicide use can be limited. Herbicides can be placed accurately at the time of planting in the space between the row pairs, after which

Paired-row planting. Morton Swanson discovered this no-till planting technique by accident. The method reduces the amount of herbicides and fertilizers required because it "feeds the wheat and starves the weeds," as one farm writer puts it. Fertilizer is "deep banded" between the close rows; the space between the pairs is mulched with residue.

an offset disk, called a composter, mulches the in-between area with residue (including weed seeds) pulled away from the drilled rows, a technique that was also discovered by chance when one of Old Yellow's disks accidentally overlapped already planted rows, removing the residue from the drilled strip. "We found," says Swanson, "that where we moved the straw away from the paired rows, they grew better."

Overall, Swanson believes herbicide use can be cut by as much as 50 percent in paired-row planting with deep-banded fertilizer placement as compared with regular no-till drilling. This obtains not only because less area has to be treated but because some weeds can't grow through the mulch made by the composter, and those that can are less vigorous than they would be under conventional fertilizer placement.

□   □   □

But the innovations—both accidental and on purpose—came about slowly. In fact, says Swanson, "We suffered for about six years. There were good crops and some very poor ones. When we were still broadcasting fertilizers we'd get into such terrible weed problems—bromes, cheat grass, quack grasses—that every so often we'd go back to conventional tillage."

Nevertheless, by 1978, Swanson was ready to offer production versions of Old Yellow and subsequent prototype drills built along the same lines for sale to other farmers. At first, the machine was called the "Pioneer" seed drill, which it certainly was, but a trade-mark-infringement complaint was brought by the Pioneer Seed Company, so the name was changed to "Yielder." Today, the Yielder No-Till Drill Company has manufacturing facilities and office head-quarters in Spokane, Washington. Yielder drills range in size from sixteen to twenty-four feet wide and in price from a bit under $100,000 to over $150,000. As of 1985, over one hundred Yielder drills had been sold, mostly to large-acreage dryland farmers like Swanson himself in a number of western states and in the prairie provinces of Canada.

□ □ □

Mort Swanson is a board member of Yielder, but he pretty much keeps to the farm and to the drawing board. The job of fabricating and selling his drill has been assigned to a group of young, enthu-siastic businessmen-agronomists who, to a person, believe their mission is as near as one can get to doing the Lord's work and still making pretty good money. Swanson's son Guy is an officer, as is Robert Papendick's son-in-law Gus Williamson. The super-sales-man is Greg Schmick, an agronomist who did graduate work at Washington State and is the closest thing to a yuppie (power-blue suit, red tie) one is liable to find in the farm-implement business, despite his rural background and even more rural clientele.

The firm is not quite so inbred as it may appear, but there is a family feeling about it, which is conveyed with a pronounced stam-mer and quite entertaining language by Schmick, who, with a port-able computer, visits prospective buyers and cheerfully cranks them out personalized Yielder spreadsheets created by a program that took over a year for Schmick and his associates to devise.

Actually, there is not just one spreadsheet but an eleven-page statistical analysis of no-till versus conventional tillage. A sample analysis, for a fictional client named Mel Moscow (Moscow being a small Idaho city not far from Spokane), has been developed to show to visiting scholars and journalists. It is based on a composite of the Yielder experience, and with it Schmick shows how farmers can increase the proportion of their acreage in active cropping, increase yield per acre, reduce costs, and increase profits using a Yielder no-till system. The figures are presented in two versions. The first is a quite conservative projection using the Yielder drill;

the second is a slightly more optimistic one using a drill called Yielder Plus. "Listen," says Schmick with characteristic enthusiasm and decibels, "if we couldn't whip anything going, we would have hung it up long ago." The figures are as follows:

	Present System	Yielder	Yielder Plus
Per-acre cost per bushel	$2.20	$1.92	$1.78
Net profit per acre	$80.99	$102.73	$128.88

On an actual cash basis, Mel Moscow's projected net profit ranges from $97,000 with his present conventional-tillage system to $123,000 with the regular Yielder drill or $155,000 with Yielder Plus. The differences are so great, says Schmick, that virtually any farmer with sufficient acreage—over a thousand, say—can buy a Yielder (dubbed the "Rolls Royce" of heavy western no-till drills by one farm magazine) on a five-year lease-purchase plan and earn as much profit, if not more, than he would using his present system, even after the lease payments are deducted. In effect, says Schmick, "the drill comes free."

Even allowing for a good dose of entrepreneurial bias in such an analysis (which is typical in the agricultural-implement industry—

Farm acreages are huge in the Palouse, as they are in most dryland farming areas—typically several thousand acres apiece. This aerial view shows a single farmstead for many square miles of Palouse farmland. Even so, with a heavy no-till drill, minimal "outside" labor is required for these large farms, and savings on tractor costs can offset the relatively high price of the drill.

no salesman would trot out a spreadsheet showing a loss), no-till farming in the drylands does have a legitimate claim to this kind of arithmetic. This is because of the impressive savings of no-till over conventional-tillage farming in man-hours and equipment hours. Schmick estimates that planting costs, for example, are reduced under no-till to one-sixth the amount required in conventional tillage, given reduced man-hour requirements. Even a huge two-square-mile farm like Mel Moscow's (or Mort Swanson's, for that matter) can be farmed by one man with just a bit of part-time help.

As for equipment, by far the most expensive item is the tractor, which in dryland agriculture is liable to be a big one, costing at least $100,000, with a useful life of about six thousand hours. According to the Mel Moscow spreadsheet, about nine hundred tractor-hours per year would be required to farm 1200 acres under conventional tillage. No-till would cut that figure to 150, meaning that a tractor could last up to six times as long—theoretically a lifetime instead of six or seven years. In flatter areas of the West, equipment savings could be further improved since a much lighter tractor would be required. In the hilly Palouse (where Mel farms), a heavy tractor is needed despite the lower power requirements normally projected for no-till versus conventional tillage.

These savings, together with improved income from increased yields—due mainly to the ability under a no-till system to plant earlier in the year, to reduce the amount of land in summer fallow, and to permit some double-cropping—are what make the case for Schmick and his colleagues.

□ □ □

Given the foregoing, it may seem a wonder that every farm in the Palouse doesn't have a descendant of Old Yellow rumbling across its fields in the spring and fall. That this hasn't happened (in fact, only about 10 percent of Palouse farmers have converted to no-till) is due to a variety of factors. An obvious one is simple resistance to change, farmers being slower in this regard than virtually all other businessmen. Another reason is lack of money. Even if you believe that Mel Moscow's profits can be yours too, it still takes a major investment in machinery to realize them, and few farmers are willing to undertake further capital indebtedness in these most difficult times for commercial agriculture. Still another reason is that no-till, or even less-demanding reduced-tillage methods, takes time to master, for a new, semipermanent soil system must be created and then farmed with the rotations and inputs that will make it work. As Carl Engle, extension specialist for Whitman County, puts it, "With no-till you can't just plow your mistakes under."

But you *can* plow root diseases under, at least to a degree—and this is perhaps the most significant factor inhibiting the adoption of no-till in the Palouse. Even under conventional tillage, various forms of root disease plague wheat farmers. But under no-till, root-rot fungi can be particularly nasty since the fungi harbored by the stubble are not broken up by cultivation, which helps to control them, but lie in wait in the undisturbed residues for the next crop to be planted. Among the culprits, pythium root rot is the most common. In addition, there is a disease terrifyingly named take-all because it "takes all" the crop. And there is Cephalosporium stripe, a fungal disease that plugs up the vascular system of the plant. There is even a no-till newcomer, previously known only in Australia, named rhizoctonia root rot. According to tests conducted by R. James Cook, a USDA plant pathologist at Washington State University, reduced-tillage practices such as no-till can cause a shortfall in wheat production of 30, 40, or even 50 percent because of root diseases like these.

Pythium is the best known of the soilborne diseases. According to one of Cook's associates, David Weller, it is present everywhere in the soil of the Pacific Northwest. A chemical fungicide is available to control pythium, but it doesn't always work. The best controls, Weller says, are long, three-year rotations—wheat, spring grain, and a legume—and turning the soil. "But these are the very things many forward-looking farmers want to get away from," says Weller. "They want shorter rotations, and they want conservation tillage in order to stop soil erosion."

Weller is working on a biological control for pythium that he believes holds much promise. The seed would be coated with a bacteria; then, as the root grows, so would the bacteria, which would produce antibiotics in sufficient quantity to inhibit the pythium fungus. In a test of the technique, Weller was able to produce a 27 percent yield increase in winter wheat. Weller believes that seed can be treated commercially for about $5 an acre—a costlier technique than chemical controls ($2–$3 per acre), but potentially more reliable.

The same biological control technique can, Weller believes, also successfully combat take-all. Take-all is a fungus that survives in the crown tissue of a wheat plant and that under conventional tillage would be broken up and its bits and pieces degraded and made harmless by soil organisms. In a no-till regime, however, the fungus is not broken up by plowing, so it can hide safely in the root crown of the wheat stubble, "just waiting for the next crop," as Weller puts it. "So when you plant that next crop right down into the crown tissue, the fungus just explodes."

Soilborne pathogens pla-
gue wheat farmers in the
Palouse and elsewhere.
Here, the dark-colored
wheat plant roots on the
left are afflicted with a fungus disease called "take-
all," so called because it
can take all of the crop.
The light-colored roots on
the right are healthy. No-till
has been implicated in an
increase in root-rot dis-
eases of various kinds, but
new cultural techniques
and biological controls
may reduce the problem.

Another soilborne pathogen is the rhizoctonia root rot, which showed up in 1984 in the Pacific Northwest. When the sample was sent to the lab at Pullman, Weller knew exactly what it was, for he had been studying the disease in Australia, where it had first been described in the 1930s. "Here is a pathogen," Weller explains, "that can grow well in soil so long as it has a food base—the straw left over from previous years in no-tilled fields. So the more you use conservation tillage, the greater the problem." Aside from chemical and biological control remedies, Weller says that rhizoctonia is ordinarily controlled by plowing since the pathogen dies out when disrupted. But a large degree of soil disturbance is not really necessary because adequate control can be achieved by "stirring the soil just a little bit."

Never mind. When the newspapers found that USDA and Washington State plant pathologists had discovered a brand-new disease in which no-till was implicated, ostensibly imported from Australia and bearing an unpronounceable name, they had a field day, further discouraging the adoption of conservation-tillage practices in the Palouse.

Weller and Cook were appalled by the outcry and by the charges from farmers that the scientists were trying to "undermine" no-till. Weller insists that the criticism is wrong as well as unfair. "We know that reduced tillage is the wave of the future in this area," Weller says, adding, "There is no inherent problem in no-till that prevents suitable yields so long as you get rid of the soilborne pathogens."

Weller and Cook are continuing their work toward this end not only by developing biological controls but also by screening various wheat varieties to find some that are resistant to fungi, especially pythium. They also stress cultural controls such as rotations, soil stirring, and the like. The point is, Weller says, "when you change one component of farming practice radically—which is what no-till does—new problems are going to arise." R. James Cook, whose international reputation as a leading plant pathologist give his words weight, stresses that no-till farmers in the Palouse really should go to a three-year rotation rather than the present two, especially for the control of Cephalosporium stripe, to give the fungus a chance to die a natural death before it can infect new seedlings. Only if winter wheat is seeded quite late in the cool fall can a two-year rotation be safe. Cook is aware of the economic implications of a three-year rotation: as the proportion of planted wheat to total crop declines, so does the borrowing power of the farmer, since wheat is the bankers' preferred collateral.

Morton Swanson appreciates the work Cook and Weller are doing but wonders aloud whether their research techniques aren't causing people to view no-till with rather more alarm than is necessary. For one thing, Swanson has found that his paired-row planting permits concentrated applications of fungicides, increasing their effectiveness and reducing cost. Moreover, on fields that have not been plowed for several years, he is having less trouble overall with root disease. He believes that a field takes more than one year to get into shape under no-till and that Cook and Weller do not account for this in their tests. In response to this criticism, Cook explains that they have been researching soilborne pathogens for twenty years. "No-till is just another wrinkle," he says, and it does not bring anything so vastly different into the picture, from the plant-pathology standpoint, that it would invalidate their methods or long-term findings.

Robert Papendick is inclined to agree with Swanson, however. It is likely, Papendick believes, that a no-tilled field tends to "equilibrate" after several years, settling down finally into steady and reliably high yields with resistance to infestations of funguses as well as to weeds and insects. He compares the disequilibrium suffered in the first years of no-till to the problems farmers have in converting to organic systems, a topic he studied as part of a USDA task force to evaluate organic farming. "Always," says Papendick, "organic farmers say that the first several years were just disastrous. But after that, they come out of it. I don't think no-till is any different. You are creating a drastic ecological change—returning the land to something much closer to its original state as native

Conservation tillage is only slowly gaining acceptance in the Palouse, but soils scientists, along with pioneer no-till farmers, are convinced that when it is more fully adopted, scenes like this will be only a memory.

grassland. Why shouldn't you expect things to happen? The trouble is, people draw premature conclusions that conservation tillage won't work and give up on it too soon. They shouldn't. Dammit. We can cut soil loss from twenty tons an acre a year around here to two tons. And we'll build our soils back in the bargain. If a hundred years ago we'd gone from the Palouse grassland straight to no-till, the organic-matter levels would probably be about the same as they were in the original ecosystem, which was so extraordinarily productive."

5

A Handful of Earth

*No landscape, however grandiose or fertile, can
express its full potential richness until it has been
given its myth by the love, works and arts of man.*
          —RENÉ DUBOS

This is the story of Carl and Rosemary Eppley of Indiana. And of
Ernest Behn of Iowa. Unlike drylander Mort Swanson, these farmers live *east* of the 100th meridian—in the most celebrated commodity-crop agricultural region in the world, the American Corn
Belt. Like the drylands, this region too has physiographical char-
acteristics that have led to the development of conservation-tillage
techniques and implements best suited to its particular crops and
soils.

The Corn Belt is a vast region taking up most of the geography
of five states, which are Ohio, Indiana, Illinois, Missouri and Iowa,
and it spills into several more—Kansas, Nebraska, South Dakota,
and Minnesota. This is the American heartland—much of it former
tall-grass prairie that, though farmed unmercifully for a hundred
years, is still the most productive of any major agro-ecosystem in
the world. Corn and soybeans, soybeans and corn, in such quantity
as the world has never seen, producing wealth unimaginable. These
are the big money-maker crops. Between the two of them, they
constitute 70 percent of farm revenues. They are, as the source of
animal feed, vegetable oil, and other products, the foundation of
the entire U.S. agricultural industry. With farm assets totaling nearly
$1 trillion, with farm products having an annual retail value of
upwards of $400 billion, and with the food business employing, one

The Corn Belt, which includes most of the Midwest, is the American heartland. It is the richest commodity-crop-producing region in the world and the cornerstone of U.S. agriculture, which is the largest segment of the national economy.

way or another, one U.S. worker out of five in the private sector, agriculture is far and away the largest segment of the American economy. And the cornerstone of this huge edifice is located in the Corn Belt.

The cropland of the region is mainly flat or gently rolling. Erosion is a serious problem, although for somewhat different reasons than in the drylands. Here, erosion from frequent, wind-driven rains often finds channels ready made in the cultivated crop rows. In Iowa, at the heart of the Corn Belt, the greatest quantity of topsoil in any state in the Union, regardless of size, is washed away every year—261 million tons, averaging about 10 tons per acre.

But there is yet another serious soil problem to contend with in the Corn Belt, a particularly troublesome condition in wet-area agriculture—that results in poor yields from changes going on underneath the surface of the ground as well as from loss of soil from the top of it. Just as dryland farmers have a vicious circle of moisture loss and erosion to contend with, farmers in humid regions have their own special kind of vicious circle. On the silty loams so prevalent east of the 100th meridian, the more you plow, the greater the compaction just beneath the plowed layer of the soil. The greater the compaction, the less the volume of soil that is available to the roots of the corn and soybean plants, thus significantly reducing their yield. And perversely, the less the yield, the more apt are farmers to plow some more in an effort that not only fails to get rid of compaction, but increases it. Research in the Midwest has shown yield reductions of as much as 60 percent (in an Illinois test plot) due to compaction, according to a report published by the Elanco Products Company, the agricultural-chemicals division of Eli Lilly and Company.

Something less severe but still troubling was happening in the Indiana cornfield just outside Carl and Rosemary Eppley's dining-room window. The Eppleys are in their sixties, with grown children. They look like a cheerful version of Grant Wood's American Gothic farm couple—he with a lean, bony frame and a face that has known the out-of-doors through the seasons, she with a motherly apron and a direct, pleasant manner. But there is nothing old-timey about the Eppleys—they are twentieth century all the way. Carl started with 6 acres and a house in the 1940s and now farms 550 acres and owns three houses. After the children were well along in school, Rosemary got a job at the local radio station and wound up as principal newscaster. Later, she worked as a farm-paper editor. Now retired from the hurly-burly of the media world, she keeps her hand in as a bookmobile volunteer. Despite the rural backdrop, the Eppleys are remarkably cosmopolitan. They sent their children abroad and took in foreign students as part of the American Field Service program. Today, son Nathan, a former Peace Corps volunteer, is an executive at a large manufacturing concern in nearby Wabash. Daughter Karen, formerly a schoolteacher and now proprietor of a dog-grooming company, is considering coming in to help manage the Eppley farm operation.

Rosemary does a good bit of this work now, but she confesses that only since the late 1970s has she concerned herself much with the science of agriculture. "I never appreciated what the soil was until I went out to the field with those men from Purdue to discuss our problem." She is referring to the cornfield outside the dining-room window. Carl had to be away that day, she says, and left instructions with her to "find out what was wrong with that field." And so the men from Purdue took their core samples and gave their verdict: compaction. What to do about it was another question— agricultural experts are typically more precise on scientific diagnosis than on practical remedies. It fell to Mel Boyer, the district conservationist of the U.S. Soil Conservation Service, to suggest that Carl and Rosemary give a new technique called "till planting" a try. And they did. First 10 acres, then 20 more, then the whole farm "went on ridges," as the phrase of art has it.

"In the spring of '79," Carl says, "the local conservation-district board acquired a Buffalo planter and cultivator, which at that time was the only piece of equipment on the market for till planting. Ernie Behn in Iowa was the pioneer, and I went up to see him. But I still had to work it out on my own here, pretty much. But that Buffalo planter came right in there and moved those stalks right to the center, planted that corn, and she never plugged once." The Eppley farm has since become a showcase of ridge-till, a form of conservation tillage especially suited for the flat, cold, and silty soils of the northern part of the Corn Belt, from Ohio straight through to eastern Nebraska. Here conventional tillage so easily compacts the soil that most fields have to be drained artificially with drain tile. The winter frost heave can break up some of the compaction, but it is quickly reestablished during the growing season by repeated cultivation trips across a field.

Ordinarily, no-till would be the answer for this condition since it would reduce trips across the field from a dozen to two or three a season. This way, the deep freezing from the gelid arctic air sweeping down from Canada could, after spring thaws, keep the soil layers open so that the fields would be able naturally to drain away the

Spring is slow to come in parts of the Corn Belt, the soil slow to warm. With ridge tillage, as practiced by Carl and Rosemary Eppley, the soil warms up faster than it would with no-till.

rainwater, which tends to "pond" on the flat, glacier-leveled fields of the Midwest. The trouble is that most of the Corn Belt warms up slowly in the spring. And without spring plowing to expose bare earth to the sun, the fields stay cold, insulated against the warming rays by the residues of the past season's crop. This organic matter, helpful in so many other respects, becomes a problem rather than a solution in much of the Corn Belt. The growing season is short enough as it is, and no-till, in this climate, makes it even shorter.

□ □ □

Ernest E. Behn, whom Carl Eppley went to visit, is a kind of latter-day Edward Faulkner. A former government soil conservationist, he took early retirement twenty years ago to conduct tillage experiments on his own farm near Boone, Iowa. Like Faulkner, Behn felt that the traditional agricultural wisdom was lacking in common sense and quit the government in frustration in order to experiment with new techniques. In Behn's case, the cause of his discontent was deep concern about the ineffectiveness of traditional remedies to stem soil erosion—contour plowing, terracing, and grassed waterways. These nostrums, prescribed by the USDA since the 1930s, just weren't working well enough to keep the topsoil in place in the Corn Belt.

Behn is like Faulkner in one other respect, too. He has written a book, its most recent edition self-published in 1982. Unlike Faulkner, however, Behn wastes little time excoriating the moldboard plow and gets right down to practical matters. If the book is less capable of attracting a widespread readership than Faulkner's *Plowman's Folly*, it is bound to pique the interest of fellow farmers. He has titled it *More Profit With Less Tillage*.

Behn points out in his book that with larger tillage equipment— rigs that plant or cultivate twelve or more rows at a time—contouring, the chief means promoted by the USDA to help stem erosion, becomes less practical since turning the equipment, even a few degrees to follow the slope of the land, leaves "dozens of small weed patches," which so infect the crops that farmers are inclined to give up contouring and just run the machinery straight ahead, slope or no slope. As for those who keep trying to contour with large, modern machinery, Behn says they must increase their use of herbicides to keep the small weed patches under control. This in turn often wipes out grassed waterways, another conservation remedy designed to carry storm water safely off the fields. With the grass killed off by herbicides, the waterways are not only useless but a positive invitation to gullying.

For these reasons, among others, "over 60 percent of the land in the United States that needs contouring is not being contoured," says Behn. And in any case, contouring is only 50 percent effective in controlling erosion in ordinary circumstances. "We must strive for 100 percent of control," Behn insists. "Not just 50."

The answer is, of course, conservation tillage—no-till or at least minimum tillage—to keep the residues on the surface and thus keep the water from running off. "Many people will be surprised to find that most of the problems of residue farming—the insects, the diseases, weed control, cold temperatures, delayed planting, poor stands, et cetera—have already been solved," claims Behn. The solution did not come so much from university specialists but from farmers like Carl and Rosemary Eppley and Behn himself, working the problems out in practice. Indeed, the specialists, Behn says, have often been a part of the problem rather than a part of the solution. They should, he says, "get in tune" and try to solve the problems of conservation tillage "instead of repeating that old, old story of cold, wet ground."

Ernest Behn's experiments began in the 1960s, with the substitution of chisel plowing for the moldboard plow. Using a chisel plow can leave as much as two-thirds of the residue on the surface. The

Carl Eppley visited the Ernest Behn farm in Iowa, shown here, to learn about "ridge till," a conservation-tillage technique more suited for colder areas, where the soil warms up slowly in the spring. Behn, a pioneer "ridger," began his conservation-tillage experiments in the early 1960s and has published an authoritative handbook on ridging.

results, in combination with the terraces Behn had already built to reduce erosion, were excellent. Yields went up 20 percent, and runoff was almost eliminated. "During 1963," Behn writes, "the terraces never carried any water! What good news this was. This meant that no insecticide, no herbicide, no phosphate or nitrate, and no silt went down the waterway to the nearby stream."

While chisel plowing was not truly "minimum" tillage (and certainly was far from being no-till), it was nevertheless an improvement over moldboarding. Still, "chiseling" was, for Behn, only a way station en route to the innovative tillage system he had in mind. "Some prefer to use a chisel and thus mix residue in the topsoil," says Behn. "This is fine, but it is not a minimum-tillage method. It takes a lot of power." Another method Behn feels falls short is the use of a heavy disk (the implement suggested by Edward Faulkner, though with reservations) to chop residues and mix them into the surface layer of the soil. And still another is the use of a rotary tiller (such as the one Louis Bromfield had success with at Malabar) to cultivate a relatively narrow strip of soil for the seedbed, leaving residue on the surface between the strips. Rotary strip tillage is especially popular in Nebraska, which has a dryer climate than much of the Corn Belt. According to the Conservation Tillage Information Center, under strip-till, as they call it, "approximately one-third of the soil surface is tilled at planting time." Behn is cautious about this technique, however. "If the soils are a little wet," he says, "this is too much tillage to suit me since it could easily damage soil structure."

During the time of his chisel-plow experiments, Behn went to visit relatives in Arizona and observed that cotton farmers there built high ridges on which the cotton was planted, with the channels between them used to carry irrigation water. On seeing these ridges, it occurred to him that a variation of this idea might solve the problem of adapting minimum tillage to the cold Iowa corn country. Permanent planting ridges raised above the surface of the field would warm up more quickly—as quickly certainly as the furrows of a plow—and between the ridges, instead of irrigation water as in Arizona, there would be space for residue aplenty, to reduce erosion and runoff and add tilth and nutrients to the soil.

Hurrying home, Behn found that, in fact, agronomists at Iowa State University and at the University of Nebraska had something of the same idea and that Fleischer Manufacturing, Inc., a Columbus, Nebraska, farm-implement company, had already built machinery—the Buffalo Till Planter—that could accomplish pretty much what Behn had in mind. That summer, says Behn (it was 1964), "for the first time, I noticed a till planter on display at the

Iowa State Fair. This machine had a sweep intended to cut off the top of a ridge and push residue aside. Behind this was a short boot to plant seed at the desired depth in the clean space just created by the sweep. Following this was a narrow press wheel with a rubber tire about three-fourths of an inch wide on it that would push down into the deep groove made by the planter shoe and press the seed down into firm, moist soil. Following this were two small disks that would cover the seed. This certainly looked like exactly what I needed. I saved money and bought one."

Behn found that the Buffalo planter permitted all the residues to remain on the surface but that planting on the high ridges— eight inches or so above field level—had none of the disadvantages of no-till, including the disadvantages of the old, old story of cold, wet ground. "The problems are all gone," Behn marveled. "I know no one will believe it, but they are all eliminated!"

Behn says that the way to get started as a "ridger" is to cultivate a second time during the summer, in whatever row crop is being raised, but instead of leaving the soil flat and level, to set the disks or ridging wings on the cultivator so that it throws a large quantity of earth—taking it from the space between rows—onto the crop row so that there is a ridge seven or eight inches high. This treatment creates an alternation of ridges and valleys across the field, which remain in place.[1] They are permanent, just as permanent as the raised beds of a backyard organic gardener. The routine thereafter is to run the till planter down the ridges you built that first summer. A narrow strip atop each ridge, ten inches wide, is cleared by the "sweep," and behind it a furrow is opened, seeded, fertilized, and closed by other tools affixed to the planter. The residue remains in the valleys between the rows, where it can hold the moisture and act as mulch against weeds. In the cleared ridge top, the crop has a chance successfully to compete with weeds. During the growing season, a cultivator, carefully set so as not to go too close to the ridges, runs down the space between the rows and uproots any weeds before they grow and spread. At the same time, the cultivator

---

[1] Like Morton Swanson and a good many other conservation-tillage pioneers, Behn too has become an inventor. He has developed the "Behn Handy Ridger" and the "Behn Ridge Mate." The former is a ridging wing attachment for the Buffalo cultivator and may be used with a "Behn Deep Shield" to protect the crops while the ridging is going on. The "Ridge Mate" is a device that can be attached to the tool bar of the popular International 800 planter. The Ridge Mate makes a high ridge and permits a minimum of "topping" with a sweep cutting down only an inch or two rather than three or four inches. "It can be wet down there," says Behn's brochure.

Close-up of "ridging wings" on a Buffalo cultivator. The ridges made in ridge tillage are permanent, something like a field version of a gardener's raised planting bed.

rebuilds the ridges as Behn describes, by setting the wings to throw the soil and residues stripped off during planting back onto the ridges.

Then you harvest—a better yield, usually—and then you're through until planting time next spring. The advantages of this method are manifold. Behn lists twenty, which may be summarized as follows.

Removing the residue from the top of the ridge during planting allows the seedbed soil to dry out and warm up early in the spring, and since the ridge top is clear, the planter does not clog up. "The tops of ridges, always moist, firm, and mellow, are a perfect seedbed with no tillage of any kind." Ridge-till allows earlier planting, even earlier than with fall plowing.

Between the ridges, the residues serve many functions, from helping to support a tractor in wet spots, to maintaining moisture, to providing shelter for wildlife. Most important, erosion is held in check. Wind erosion is simply no longer a problem, and water erosion is cut to a minimum. During heavy rains, says Behn, the water collects in the valleys, which are filled with last year's residue, thus inhibiting runoff.

"Only two tillage operations are necessary to harvest time," says Behn. The first is for planting, the second for cultivating and rebuilding the ridges. Since the tractor wheels never run on the ridges but only between them, compaction in the root zone of the plants is greatly reduced.

Equipment requirements are greatly reduced. As Behn puts it, "No plow, no disk, no harrow, no rotary hoe, no chisel, no field cultivator, no spring tooth—but just a planter, a cultivator, and stalk chopper." In addition, fuel costs can be cut 50 percent since these implements need not be pulled across the fields.

According to Behn, this system is "environmentally sound and in close harmony with nature." Weed control is easier since weeds are pushed to the valleys during planting and then are easily taken out by the cultivator. In fact, he says, "herbicide usage is reduced and may eventually be eliminated. This is not proven, but it is logical."

And, finally, the bottom line: "Yields are increased or at least maintained as compared to those of conventional farming."

□   □   □

These were the features of ridge tillage that, in the late 1970s, convinced Carl Eppley, on the advice of conservationist Mel Boyer, to give the technique a try himself. But to check it out thoroughly, for the first three years Carl kept careful records of differences in yields in a side-by-side comparison of corn yields under ridge tillage and under conventional tillage on two adjacent ten-acre plots. The results were as follows.

Corn Yield (Bushels per Acre)

	Conventional Tillage	Ridge Tillage
1979	149	156
1980	122	145
1981	142	155
Three-year average	137	152

Soybeans also improved in yield, though not so dramatically. While Eppley did not conduct a side-by-side test for beans, his overall yield results confirm the corn-yield test findings.

Soybean Yield (Bushels per Acre)

1975–1980	(Conventional tillage)	46
1980–1985	(Ridge tillage)	52

Unlike no-till, ridge tillage requires some mechanical cultivation for weed control rather than relying totally on herbicides. The cultivator here uproots weeds between the rows but leaves the planted ridges intact.

In addition to these increases in yield, Carl and Rosemary have saved money on labor and other input costs—to the tune of $40 to $60 per acre.[2] In the period just prior to changing over from conventional to ridge tillage, the Eppleys employed three part-time laborers during the season (earlier, they had a full-time employee). Now, virtually all outside labor costs have been eliminated. Carl does all the planting and cultivating himself and hires someone to help for a week or so during harvest. Fuel use has been reduced by two-thirds, and equipment maintenance and replacement costs are greatly reduced.

While as a general rule the various forms of conservation tillage require greater use of herbicides to make up for reduced mechanical tillage, this is not necessarily the case with ridge-till. One practitioner from Minnesota wrote Ernest Behn that not only did he realize a 50 percent reduction in labor and a 30 percent reduction in machinery overhead, but he also saved on herbicides. "The sys-

---

[2]It is important not to miss the significance of this saving. A $50-per-acre decrease in costs drops straight to the bottom line. On a five-hundred-acre farm, the total would amount to $25,000 a year. For the farm family, this is the same as real income, as if it were an increase in salary. A $25,000 raise is no small amount in any middle-class family budget, whether on the farm or in the suburbs. Farmers understand better than most the truth of the saying, "It's not what you earn, it's what you keep."

tem provides better weed control with less chemicals used," he said. Behn agrees: "I never use more herbicide than good conventional farmers use, and I frequently use much less." He believes that most weed problems with ridge-till come about because of poor timing in cultivating rather than anything inherent in the system. "I am convinced," he writes, "that a man giving full attention to the problem of weed control in a ridged minimum tillage field will be just as successful as the conventional farmer, without extra herbicide costs." In fact, these costs seem to decline over time for the ridge tiller; at least this has been Behn's experience as well as the Eppleys'. This is because buried weed seeds are slowly eliminated (they are no longer turned up by plowing) and fewer new weed seeds are added. What weeds aren't taken care of by the mulching action of the residue between the ridges and by crop competition on top of the crowns are cultivated out mechanically during the growing season before they have time to come to seed. Weeds that might have emerged on the ridges before spring planting are cut off and swept into the spaces between the rows, where a heavy mulch from the previous harvest eliminates early weed growth. During the growing season, the cultivation runs take care of any weeds between the rows, throwing the mulch and soil back onto the ridges, which inhibits weed growth in the crop rows. Although there is less mulch in the "valleys" after midseason cultivation, by that time most weeds likely to emerge before harvest will have already germinated.

This is not to say that herbicides are not an important element of successful ridge tillage. Carl Eppley regularly uses Bicep, Banvel, Basagran, Dual, Lexone, and even Paraquat—but in smaller amounts than in no-till. For corn, herbicides cost him about $11 an acre and for soybeans, about $14. This is a typical cost for Indiana, according to Purdue University studies, which show no difference in per-acre cost for herbicides for conventional (moldboard) tillage, chisel plowing, and ridge tillage. No-till herbicide costs are about 50 percent higher.

The actual field experience gained in controlling weeds with carefully applied herbicides backed up by mechanical means in the systems developed by innovators like Ernest Behn has pursuaded many midwestern farmers to adopt ridge tillage, just as the Eppleys did. In the Corn Belt, about 1.5 million acres of cropland were ridge-tilled as of 1985. According to Purdue soil scientist William Moldenhauer, Corn Belt farmers using conservation tillage are becoming quite confident that they can control weeds at a reasonable cost—despite, as Moldenhauer puts it, "the dire early predictions of weed scientists who cautioned that with conservation tillage you can build up a weed problem that you can't handle." The fact is, not only can farmers like Carl Eppley handle the problem, the problem seems to be diminishing in severity as the years pass.

In Carl Eppley's case, if not that of other ridge tillers, not only is it possible to reduce herbicide use, but also very little insecticide is needed. The corn-bean rotation usually keeps serious infestations from developing. With the permanent ridges—the raised beds upon which no traffic is ever allowed—compaction has been alleviated. The soil is building up so much tilth that it is carrying over more nitrogen from year to year than had been the case before, which someday may result in significantly reduced fertilizer costs—a result that Edward Faulkner had predicted for "plowless farming" but that had been scornfully dismissed by agricultural scientists. Now, forty years after Faulkner, Eppley is beginning to wonder if he and Rosemary aren't creeping up on some form of organic farming. "The soil is getting to be like a garden, just as mealy and nice," says Carl. "And the fish worms are coming back," Rosemary adds.

□   □   □

Another thing that has come back to the Eppleys' farm because of conservation tillage is peace of mind. Suddenly the high-pressure demands of traditional farming have been reduced, especially the "hurry-hurry-hurry" of fall plowing and spring planting, as Ernest

Carl and Rosemary taking a break. The Eppleys have found that ridge tillage has reduced much of the pressure of farming; there's more time for hobbies and each other. They are popular lecturers in the Midwest on ridging.

Behn puts it. With no fall plowing required, and trips across the field reduced by two-thirds or more, a ridge-till farmer succeeds by applying his brains to the practice of agriculture rather than his backside to a tractor seat. Planting in the spring is a much more flexible activity for the ridger, for he can get into the fields earlier than his conventional-tillage counterpart. He need not be concerned about the clodding that obtains when fields are plowed too early in the wet springs of the Corn Belt. For the ridger, all it takes is one pass to sweep the top of the ridge and plant the seed, as opposed to the several needed by the plowman, who must also harrow the surface once or twice to get a proper seedbed.

During the growing season, one or two quick cultivating trips are all that is usually needed until harvest. And after harvest (which is often earlier than the conventional farmer's), the job is done. No fall plowing; just pack it in until the following spring—a clear stretch of six months, from mid-October to mid-May. All in all, the time afield is very modest for Carl Eppley's 550 acres, a typical medium-sized farm. On such acreage, a ridge tiller can easily handle 60 acres a day, meaning that for each operation, less than ten workdays are required. For stalk chopping, planting, cultivating, spraying, and harvesting, a total of forty days or so would be required—eight work weeks. The rest of the year is scarcely discretionary for farm families like Carl and Rosemary Eppley, but the great pressures are gone.

Before ridge tillage came into their lives, says Rosemary, they, like other farm families, were "victims of stress," particularly during the planting season. "Failed communications," as Rosemary describes it. "Tired minds and bodies. Explosive emotions. But with ridge-till, we have been able to solve many of our problems—excessive expenses, insufficient time for necessities. With ridge-till, farming gives us pleasure. There's time to check the fields more often. We've improved our dispositions. There is a return of wildlife to the field because of the life-giving residue."

For Rosemary herself, ridge tillage means that she does not need to have an outside job to fill out farm income—in effect, to pay for hired hands. Instead of doing desk work late into the evening, it can be done during the day, with Carl helping. "We work together at that desk now," says Rosemary. And when the work is done, new hobbies can enter the picture. The small pleasures are, perhaps, the most important. "It was really good to sit down at the close of the day," says Rosemary. "And to have regular meals."

These are benefits not measured by the agricultural scientists at Purdue and elsewhere. But in the end, they may be equal in importance to yields or profits. What good are bushels and dollars when the getting of them destroys the promise of a good life on the land as surely as it destroys the tilth of the soil?

"Have you ever picked up a handful of earth," asks Rosemary Eppley, "and considered what you were holding? Within your hand is life. Just as individuals are stressed, and families are stressed, so too is the land being stressed by misuse. Don't you think it's time we stopped abusing ourselves and our land? After all, there's nothing quite like working together and with the land: planting, nurturing, preserving, and using this good earth, making this world a better place to live, and enjoying each other."

So saying, Rosemary looks over at Carl. He gives her a grin and nods his head in assent. And the cornfield outside the dining-room window is doing just fine, too.

6

The Enemy Beneath

This is a plea for living earth,—a
plea for life itself beyond tomorrow.
—ANN AND SAM ORDWAY

On March 12 and 13, 1986, in Pike County, southeastern Alabama, there fell the heaviest rain, 7.1 inches, ever recorded in a twenty-four-hour period by Jerrell Harden's uncle's meteorological rain meter—which has been scrupulously consulted and findings recorded every time a drop has fallen during the last ten years in these parts. This was a rain to remember. "Cats and dogs" does not describe it. To be out in it was like standing directly under Victoria Falls. To take a breath in it was to risk drowning. It was a gully washer. It could have been a disaster for the Harden farm, or anyone's. "Seven point one inches," says Jerrell Harden, stressing every word, with a tone of true amazement in his voice. Harden is a thin, somewhat ascetic and studious-looking man in his mid-fifties, and he has taken a visitor on the day following the big rain to a large, slopey field planted with winter rye.

"Look there," he demands, pointing to a roadside ditch beside the field. "What do you see?" In fact, it was an ordinary ditch— weedy, still a bit soggy from the recent rain. Nothing more. "You don't see anything, do you? No eroded soil. Nothing. And the field. See any gullies? No, you don't. No gullies. Hardest rain in ten years, maybe twenty. No erosion. None." There is a pause, because Jerrell Harden is perhaps not so used to lecturing visitors as he might wish, for he has become a well-known farmer hereabouts, through-

Heavy spring rains, typical of the Southeast, can wash away a newly planted crop in a matter of hours. This is exactly what did *not* happen to Jerrell Harden's Alabama fields despite the heaviest rainfall in ten years. Ironically, heavy spring rains like this can be followed by a disastrous summer drought, as was the case in 1986.

out the Southeast in fact, with a public. He is also well on his way to substantial wealth. This is primarily because he has figured out a way to adapt no-till farming—a technique that as a conservationist, he very much wanted to employ on his farm—to soils that were not well suited for it, soils that would compact not only from conventional plowing or disking but from just running harvesting equipment across a field on a damp day.

Coastal Plain Alabama is, you see, just about the hardest hardpan country there is in the United States; it always has been and always will be. Compaction is the farmer's "enemy beneath" that not only limits yields by limiting root growth and making the soil either too wet or too dry for proper plant development but in fact compounds this damage by being a primary cause of erosion. Compaction creates erosion because rainwater that cannot sink into the ground is rainwater that flows across the surface of it, taking topsoil and nutrients with it and gullying the land out in the process. If the rainwater can sink in, there is no erosion. This is the lesson of Jerrell Harden's slopey field of winter rye after 7.1 inches of rain in twenty-four hours. The water sank in.[1]

□   □   □

[1]As it turned out, the water absorbed by Harden's fields was more important than either he or his visitor realized that spring day of 1986, for the summer was to be the dryest on record for the Southeast, resulting in the loss of $2 billion in farm sales for the region.

You will remember that what happened to Carl and Rosemary Eppley's Indiana cornfield five hundred miles north of Pike County was compaction, too, but this was something new to the Midwest. In fact, the Corn Belt, of all the major agricultural regions of the United States, was the last to experience "the enemy beneath"— compaction zones in the soil created by the heavy machinery and elaborate regimes of postwar industrial agriculture, which require traversing a field not only for preparing the soil and planting it but for cultivating, spraying, and harvesting. But Corn Belt soils, silty loams that contain a fair amount of organic material, can withstand the pressure of farm machinery better than those of yet another great agricultural region of the United States, the Coastal Plain of the American Southeast, of which Pike County, Alabama, is typical.

There are, in fact, two problems created by compaction. The first is the effect of the hardpan itself. The pan can be created by almost any implement. Historically, the moldboard plow is its chief cause, making a densified stratum of earth just beneath the plow sole. But disk harrows too can exert a great deal of downward pressure on the soil. After a pan is formed, it is not entirely impenetrable by crop roots, but nearly so, prohibiting the roots from penetrating to

The "enemy beneath" is compaction, being measured here with a special instrument. Compaction is created by wheel traffic as well as by the moldboard plow and heavy disks. A nearly impermeable "plow pan" is clearly shown in this cutaway view. Moisture and plant roots have difficulty penetrating this extremely dense stratum.

the more porous subsoil below, which contains moisture and essential nutrients. Experts say roots that might normally grow a quarter inch a day can be slowed down to a quarter inch a *month* once they hit the compacted pan layer. The result is stunted growth early in the season—the very time when vigorous growth is needed to create crop competition for weeds.

The other dire effect of compaction on soil is the elimination of "pore space" in the root zone. An ideal growing medium would have 50 percent pore space. Compacted soils can lose half of that. The effect is that space for water retention is greatly reduced, as is the ability of the soil to provide oxygen and other gases needed by plant roots for good growth. A good deal of hybridizing has been done to create cultivars with roots strong enough to penetrate a hardpan. But this, by itself, does not eliminate the effects of compaction on plant growth. What is needed is pore space to permit "gaseous

The southeastern Coastal Plain, although a major agricultural region, has relatively poor soils, resulting in low yields for principal crops such as cotton and peanuts. As the richer piedmont soils were farmed out, row-crop agriculture shifted to the sandy, red, highly compactible Coastal Plain soils, which were formerly used only for pasture.

exchange" to the roots. Without it, a kind of asphyxiation takes place.

The combination of these two problems—the physical barrier to root growth created by hardpan and the reduction of pore space— can, as Carl and Rosemary Eppley's cornfield demonstrated, reduce yields significantly even in the more organic soils of the Midwest, which are relatively resistant to compaction. In the sandy soils of the eastern and southern Coastal Plain, the problem of compaction is even worse and has been endemic for a hundred years, resulting in extremely low yields for all principal crops, such as cotton and peanuts, and producing a traditionally impoverished farm economy and rural culture along with it.

As a physiographical region, the Coastal Plain is a band of land running from New Jersey down and across the southern United States through Texas and into Mexico. It is narrow at the northeastern end and widens as it spreads south. West of the Appalachians, which terminate in Georgia, the Coastal Plain broadens out, taking in all the Deep South and reaching as far north as the tip of southern Illinois. At first disdained by the earliest settlers, who preferred the rich, brown soils of the Piedmont, the Coastal Plain has now been brought into row-crop agriculture. Today, much of the Appalachian piedmont in the South is permanently lost as cropland, its topsoils having eroded away long ago. The old fields have reverted to pine forest, and the inherently less-fertile soils of the Coastal Plain are now used for commodity crops.

There are, to be sure, some famous, highly productive agricultural landscapes within the Coastal Plain physiographical region, albeit uncharacteristic of the region as a whole—the deep, fertile alluvial soils of the Mississippi Delta in parts of Arkansas, Mississippi and Lousiana, for example; or the famed black belt of central Alabama, a cotton-growing region that was once the bottom of an inland sea trapped behind a cuesta, which is a ridge with a long, gentle slope on the ocean side and a steeper slope on the inland side. But for the most part, the soils of the Coastal Plain are sandy, without much organic matter, and therefore extremely compactible. In these gritty, red soils, huge crops of peanuts and cotton are grown nevertheless, along with the ubiquitous corn and soybeans, wheat and other grains, and vegetables in the warmest areas.

□   □   □

Jerrell Harden got his red-dirt Pike County Coastal Plain farm from his father, who got it from his father, who got it from his father, who got it from Billy Harden, who was the original homesteader, the first of the five generations of Hardens on this land. Now the farm is 900 acres (currently operated by Jerrell, his father, and his son; brother Leo is on leave), and the family works 350 more, which are rented. The farm is near Banks, Alabama, which is near Troy, the Pike County seat and the location of Troy State University, from which J. C. Harden, Jerrell's father, graduated in the class of '83 at the age of seventy-eight, with a bachelor's degree in history. "Plow deep while sluggards sleep," says J. C., quoting Poor Richard's Almanac. J. C., one of the founders of the Farm Bureau Insurance Company, is as enterprising as Ben Franklin could wish. But that sort of thing runs in the family. Hughey Harden, J. C.'s uncle, went to Washington to shake the hand of President William Howard Taft on December 7, 1910, because he had won first prize as a "boy farmer" by producing 132 bushels of corn for $46.90 an acre. "A splendid boy," enthused the Troy *Messenger*, bestowing the sobriquet "the boy who made Banks famous." And now Jerrell Harden, who harvests 196 bushels of corn per acre with a unique tillage system he has invented, has made Banks famous all over again.

It all started in the spring of 1971, Harden recalls, when the Harden farm played host to a field day organized by the Chevron Corporation, makers of Ortho agricultural chemicals; the Allis-Chalmers Corporation, a farm-implement company that had come out with a no-till planter; and the Soil Conservation Service of the U.S. Department of Agriculture. The idea of the field day was to demonstrate to farmers in southern Alabama, and anyone else who wanted to come, how no-till worked—this new concept of "burning down" the weeds with a herbicide and planting directly into the residues without plowing the land at all. No-till would, they believed, help stem the terrible erosion that came every spring to the southeastern states, where heavy rains would perversely pour from the heavens at the very time when the soil was most vulnerable, right after spring planting in the freshly plowed fields.

The trouble was that for all its promise, the resulting stand in the Harden's no-till test field was poor. Except, oddly, at one corner of it. It was dry that year, Harden remembers, and he had "subsoiled" a part of the field used for the no-till demonstration. Subsoiling is a way to break up hardpan by the use of a heavy, inch-wide steel shank with a forward curve like an elephant's tusk, which digs deep into the soil—twenty inches or so. When drawn by a tractor, the shank, called a "subsoiler," opens the way for water and roots to penetrate through the otherwise impenetrably compacted hardpan to the looser subsoil beneath, which contains moisture, nutrients such as nitrogen, and beneficial trace elements not otherwise available to the plant's roots. "We left the trash on top in all of the demonstration field," says Harden, describing the key feature of no-till. "But where I had subsoiled, we got dramatically improved

One of the first no-till planters, manufactured by Allis-Chalmers in the mid-1960s. It was from a demonstration of this machine on his farm that Jerrell Harden discovered a way to adapt conservation-tillage techniques to Coastal Plain soils. Not because no-till worked, but because it didn't.

yields compared with the no-till area." At that point, the lights went on in Harden's brain: "I thought to myself, now that's the way to farm. I'll just get a no-till planter with a subsoiler on it." But a search for such an implement turned up nothing. "Nobody made one," says Harden, confirming his suspicions. "So I went into the shop and made one myself."

The result of this effort was an implement that Harden now calls a "Ro-Till" because it "tills the row" rather than the entire field. The space between the rows is undisturbed soil, which can remain well mulched with residues. "No-till turned out not to be a viable alternative to conventional tillage in this area," Harden says. "Our land doesn't have as much humus as other parts of the country where no-till is used. Less than one percent organic matter. It has more sand and compacts easier—so much compaction that hardly anybody can use no-till at all."

As a practical matter, then, the conservation-tillage alternatives to conventional tillage on such soils are "mulch" tillage, using a chisel plow and leaving as much residue on the surface as possible, or Harden's Ro-Till. According to Ken Rogers of the Soil Conser-

vation Service in Auburn, even mulch tillage is difficult in southern
Alabama. In those parts, he says, "Harden's is about the only kind
of conservation tillage there is." Harden himself supplies the reason: "Chiseling might be deeper than a moldboard or a disk," he
says, "but it still doesn't break the hardpan." In Harden's view, good
conservation farming in compactible soils requires not only that
residue be left on the surface of the field but that openings in the
hardpan be made so that the roots of crop plants can go deep and
rainwater can recharge subsurface strata rather than running off
and causing erosion. Subsoiling was the reason why that slopey field
planted in winter rye eroded not one bit after 7.1 inches of rain.

It took Harden two years to perfect the first version of his implement and, joining forces with Brown Manufacturing Company, a
small firm in nearby Ozark, to offer it for sale in the Southeast,
where Brown has distribution capability. But the Ro-Till became
so popular that it became clear that more aggressive marketing
and wider distribution than Brown could provide would pay off.
Accordingly, Harden cut a six-figure deal (plus royalties) with Bush
Hog, a division of the giant Allied Products Corporation and one of
the largest implement manufacturers in the country. Bush Hog,
located in Selma, Alabama, insisted that Jerrell Harden's brother
Leo agree to be product manager for Ro-Till, which was fine all
around. Leo Harden says that the Ro-Till is top priority at Bush
Hog, as indeed it might well be. Within a few months after its
introduction by Bush Hog in 1985, orders for the Ro-Till exceeded
production. By 1986, sales had been chalked up in thirty-three states,
with an aggregate (including the earlier Brown version) of over
fourteen hundred units sold.

The Ro-Till can be used for ridge-till as well as strip-till. The
strip tillage provided by the "in-row chisel," which is the generic
name of the Ro-Till (and increasingly useful, since Harden's invention now has its imitators) is quite different from a kind of conservation tillage used in Nebraska and some other Plains states, which
is also called "strip-till" by the Conservation Tillage Information
Center and some others. This kind of strip-till uses a rotary tiller
(like a garden rototiller) to prepare a planting bed by mixing residues into a narrow strip of soil to a depth of two or three inches,
leaving the residues in the spaces between the rows undisturbed.
Although both techniques limit tillage to a "strip," the use of the
same term for both the rotary-tilled strip and the in-row-chiseled
strip is misleading and confuses many people.

Variations in the settings of the disks of Harden's Ro-Till can
produce a level planting strip; one that is slightly lower at the center, useful for certain crops; or one that is raised substantially higher

Jerrell Harden and a close-up of the Ro-Till he invented. The implements, from front to back, are a coulter to cut the stubble, the subsoiler shank, fluted disks to help pulverize the soil, and a rolling basket to incorporate dry herbicides.

as a permanent planting ridge. Both Jerrell and Leo agree that because of the versatility of the Ro-Till for a number of different conservation-tillage systems used in various parts of the country and its relatively modest cost (a four-row unit with a planter attachment is $8,900; without, $8,100) compared with other tillage equipment, the Bush Hog sales goal for the Ro-Till of $7 million in 1987 may not be overly optimistic, despite the grim financial situation now facing American farmers, especially in the South.

Bush Hog now has a first-rate, four-color, twenty-minute promotional videotape for the Ro-Till (produced by the Modern Communications Group in Mobile), replete with an Orson Welles-type voice-over, beauty shots of lush farm fields, classy opticals, and swelling music. The star of this production is not Jerrell Harden, however (he has a supporting role), but Albert Trouse, Jr., a highly respected soils scientist, now retired from his post at Auburn University's National Soil Dynamics Laboratory. (The name was recently changed from National Tillage Equipment Laboratory.) While at

Auburn, Trouse helped Harden all through the development of the Ro-Till and later signed on as a consultant to Bush Hog. It may take a bit of imagination on the reader's part, but Trouse's on-camera description of the workings of the Ro-Till on the Bush Hog tape is as good a way as any to explain the complex way it creates a crop-growing zone twelve inches wide and twenty inches deep ideally suited for areas like the southeastern Coastal Plain with highly compactible soils:

> First [Trouse begins], we have a coulter [points to coulter in the front of the unit] that works down about three or four inches to cut the trash [heavy stubble is shown] so it won't wrap around the subsoiler shank. . . . Now, you notice the subsoiler point will sheer up at about a forty-five-degree angle ahead [like the elephant's tusk—the upward thrust reduces the chances that the point of the subsoiler will, like the moldboard plow it replaces, create its own plow pan]. Notice the height of the soil in front [it is a bit higher than the surrounding soil, like the bow wave of a ship] and then how much it is loosened behind. So now we have much looser soil; we don't want that firm soil [points to undisturbed surface between the rows] to grow the crop in. We want to pierce the plow pan. Now the subsoiler throws the soil ahead and out to the side. These disks here [points to a paired set located behind the shank] bring it back toward the seedbed, and they're fluted to help pulverize the soil. [Adopts professorial tone] Subsoiler throws it out; these [points to fluted coulters again] bring it back and pulverize it. And then to complete the pulverization, to make a decent seedbed, there's a rolling basket [points to cylindrical device that looks a bit like the reel from an old-fashioned lawnmower] directly behind that makes a nice pulverization to make a good seedbed. You notice [points at the tilled strip] how clear it is here of debris, so the seed can be put into the proper depth [tight shot of seedbed].

After a brief scene with Harden saying that the rolling basket can also incorporate a herbicide, the lack of which, he points out, has heretofore limited the adoption of conservation tillage, the camera returns to Trouse, who is standing in a trench to demonstrate, via a cutaway view, how the Ro-Till operates *underneath* the surface of the ground:

> What we have here [says Trouse] is a new principle. It's been introduced under the name PAT, precision applied tillage. The only tillage done is right in the row, to prepare a real good production zone. It's supposed to be deep enough to pierce these barriers [indicates a ledgelike sheet of compacted soil] so the roots can go down and water can be funneled down. Now here you have an example of one [points to a cross section of a Ro-Tilled row] that has no plant in it. You can

Albert Trouse, a soils scientist, demonstrates the Y-shaped production zone created by the Ro-Till subsoiler. The shank opens a way through the hardpan for water and roots to reach the subsoil.

see the "Y" that the subsoiler makes; it has pierced this pan; rain-water that comes down in the spring and summer will funnel down through this "Y" and get into this very permeable subsoil. This soil [indicates subsoil beneath the hardpan ledge] is extremely loose. It's a production zone like you don't have in conventional tillage, where all the roots and water have to go through these compacted layers. With this [points to the "Y" again], the water and roots can zoom down at an extremely fast rate and get into the subsoil and exploit this reservoir real early. All the summer rains will come down and recharge this reservoir. There's almost no runoff. And in the mean-time, if you don't cultivate [indicates the untilled, mulched area between the rows], you have a good, solid traffic zone.

□   □   □

The solid traffic zone to which Trouse referred is a critical issue in the technology of modern tillage on compactible soils. The cruel irony of compaction is that, in general, the more you try to get rid of it, the worse it becomes. According to James H. Taylor, an agri-cultural engineer at Auburn's Soil Dynamics Lab, soil is at its most vulnerable when it loses its "structure." And the loss of structure takes place on two occasions: when the ground is wet (as in the

spring, often during planting, as well as in the fall, during harvesting) and after it has been plowed. And the deeper the plowing,
the deeper the compaction. This is an important aspect of the basic
paradox of tillage, which is, as Taylor puts it, that "most of the
tillage we do is to remove the effects of wheel traffic. If we never
put a wheel in the field, we'd seldom need to put a plow in the
field."

Therefore, says Taylor, subsoilers, especially if used indiscrimi-
nately (which is not necessarily the case with the Ro-Till, but nevertheless a problem), exacerbate compaction. It takes a big tractor to
pull a subsoiler, Taylor says: thirty horsepower per shank; more if
other implements are pulled along with it, such as the various disks
and wheels of a planter. For a typical four-shank subsoiler, then, at
least 130 horsepower would be required for the tractor. Such a
tractor would weigh in the neighborhood of six tons.

There is yet another paradox beyond this one, which is that when
a field is plowed to make it "perfect" for seeds to germinate efficiently and plants to grow, its surface is the most inefficient possible for tractor tires as well as those of other implements, which
need a solid, "tractive" surface, not a soft one. Accordingly, the more
plowing, the softer the surface; the softer the surface, the more
horsepower is needed for a tractor to pull through it. And the

James Taylor, an agricul-
tural-equipment engineer,
measures compaction
caused by a tractor tire. "If
we never put a wheel in
the field," says Taylor,
"we'd seldom need to put
a plow in the field."

more horsepower of the tractor, the greater the weight; the greater the weight, the deeper the compaction in the soft soils; the deeper the compaction, the deeper the subsoilers must go; the deeper they go, the softer the soil for traction the next time, leading to more compaction. And so on.

This problem has become so widespread, says Taylor, that no agricultural region can escape the perils of compaction. Indeed, he says, "A lot of things we once thought were nutrition problems in the growth of crops are actually compaction problems. Some people think that freezing and thawing cures everything. But now, with this very heavy machinery, compaction goes so deep into the soil that it is well below the frost line. Some of the grain buggies have eighty thousand pounds on one axle. That's probably making the Chinese squeak, it's compacting so far down."

No-till, says Taylor, is not a complete answer for this dilemma because even though there is no loss of soil strength from plowing, the cumulative compaction of harvesting implements (like the "grain buggies") still takes place. Such compaction can be extremely severe when the soil is wet, as it often is during harvesting operations. In no-till, moreover, compaction can obtain on the surface as well as deep underground. The surface compaction reduces water absorption. While residues keep the water from running off quickly and causing erosion, the water still runs off. "I don't think we can interrupt the compaction cycle just by stopping tillage," says Taylor. "We've got to do something about traffic at the same time."

□   □   □

As any backyard vegetable gardener knows, one way to reduce compaction amongst the pole beans, carrots, and zucchini is to make permanent paths in the garden and never to walk in the beds, but if you must, at least to put down planks between the rows so that the weight of the gardener will be distributed over a wider area of ground than the soil directly under his or her footprint. As it happens, these are the exact strategies proposed by James Taylor to deal with the problem of compaction in commercial commodity-crop agriculture. His theories apply to all kinds of tillage but may be especially valuable in the evolution of conservation tillage.

Taylor's idea is to separate the "cropping zones" from the "traffic lanes." He calls the concept "controlled traffic," and since 1979, when he was named national research leader on this concept by the USDA's Agricultural Research Service, it has become the most compelling professional preoccupation of his whole career. A pleasant, soft-voiced man, Taylor quietly but firmly insists that, given today's heavy machinery, this separation must eventually come about if no-till and other forms of conservation tillage, including Harden's Ro-Till or some variation on it, are to fulfill their promise.

The ultimate in controlled traffic, Taylor says, is to dedicate permanent traffic lanes in the field, as widely separated as possible. And this can be extremely wide. An experimental machine built in California has wheels thirty-three feet apart, leaving a "cropping zone" as wide as a city street. At this kind of distance, quite permanent lanes would become feasible. Given a permanent roadway to bear its weight, such a machine could till, plant, cultivate, and maybe even harvest a dozen crop rows at a time with near-zero compaction in the cropping area. Moreover, because tractive efficiency in the traffic lanes would be increased enormously, horsepower requirements could be reduced along with fuel needs. Says Taylor, "Each one-percent increase in tractive efficiency brings savings of at least forty million gallons of fuel annually consumed by United States farmers."

Taylor is the first to understand that such a radical departure in tillage machinery is not likely to happen soon, and he stresses that the prototype machines developed so far are solely for experimental use. "For research work," says Taylor, "the 'spanner' is important, but it's not practical now for field machinery." What *are* practical, Taylor believes, are some modifications in certain conservation-tillage practices that can produce some of the beneficial effects of

An experimental "spanner," demonstrating how dedicated traffic lanes can be separated from cropping zones in the agriculture of the future. Certain kinds of conservation tillage already produce some of the beneficial effects of reduced compaction possible with a spanner, but on a more modest scale.

the "spanner" on a more modest scale. He is particularly interested in adaptations of ridge tillage, which would pertain to Harden's Ro-Till as well as to more traditional ridging equipment such as the Buffalo equipment used on the Eppleys' farm in Indiana. "Since controlled traffic is a cropping system in which the traffic lanes and crop areas are distinctly and permanently separate, ridge tillage, by its very nature, fits that definition," says Taylor. "In the beginning of ridge tillage, farmers had to put a wheel in every furrow. The equipment was not made to eliminate this. But we are now making great progress in ridge-till by pulling more and more of those wheels out of there and spacing them farther apart so that we are concentrating our traffic in a couple of furrows every ten or twelve feet. But even if you do have a wheel in every furrow, you fit the definition of controlled traffic. The key is to keep the same ridges. The progress I hope to see in ridge-till is to reduce compaction by eliminating more and more of the wheel traffic between the ridges."

Taylor points out that when it comes to compaction, axle weight is the basic problem. "Many farmers use dual wheels on their tractors to take care of the compaction by spreading the load across a larger surface of the ground," he says. "But in most cases, they're actually just doubling the compacted area. The reason for this is that it's the first pass, when the soil is soft, that does ninety percent of the damage." To reduce this, Taylor has been studying different wheel shapes. "What's important," he says, "is the footprint." And the most beneficial footprint of all, when considerations of traction, load distribution, and soil dynamics are all taken into account, is not a wheel at all, no matter how big around it may be, but a pneumatic track—a cross between a tire and a continuous steel track of the kind used on bulldozers and tanks. (Steel tracks were also used, in the old days, on farm tractors. They were replaced by heavy-treaded rubber tires because tires permitted easier over-the-road travel.) Driven by differentials on both front and rear axles, the pneumatic track, given a flexible radial-bias design, can have as much versatility as a round tire but can increase tractive efficiency enormously while reducing compaction because of the much longer "footprint," which permits a narrower profile and a commensurate reduction of the surface area bearing the weight of the machinery. If, as suggested at the outset, the dedicated traffic lanes in a field were analogous to the permanent pathways in the backyard vegetable garden, then the pneumatic track would be analogous to putting a plank across the surface of the garden to walk on instead of sinking into the soil at each step and compacting it.

□   □   □

While neither the "spanner" nor the pneumatic track is quite so Buck Rogerish as it might appear (Taylor believes some of these new technologies might well be applied in highly intensive vege-table-crop agriculture fairly soon), their practical application is limited even under the best of circumstances provided by conser-vation tillage. In fact, Jerrell Harden says that he does not neces-sarily keep to given traffic lanes in his own fields from year to year, and for a very practical reason: in his rotations, some crops must be planted thirty inches apart and others, like cotton, thirty-six. Even in corn and soybean rotations in areas where ridge tilling takes place, new narrow-row cultural methods for soybeans are taking hold that do not match the row-width requirements for corn. This means that conservation-tillage farmers would either have to give up ridging, substituting chisel plowing, most likely, or give up rotations, meaning that they would lose the advantages, such as insect control, that rotation provides.

In most cases, therefore, the likelihood is strong, except in quite advanced ridge-tillage techniques, that wheel pressure will con-tinue to be applied fairly generally across the fields of even the most dedicated conservationists. This exigency set Charles B. Elkins, a

Charles Elkins demonstrating a variation on the Harden Ro-Till. The subsoiler shank is shortened and a thin blade is substituted to make a tiny opening in the subsoil so that the hardpan can retain its ability to distribute the weight of heavy machinery, eliminating deep subsoil compaction.

colleague of Taylor's at the Soil Dynamics Lab, to wondering if some sort of accommodation might be possible with hardpan. "At thirty horsepower per shank," says Elkins, referring to the deep subsoilers, "it takes a lot of energy to rip up a plow pan only to recompact it later from wheel pressure." And what is worse, the postsubsoiling compaction would go deeper than before. "I think," says Elkins, "we ought to figure out a way to leave the plow pan in place to guard against the compaction of the loose subsoil."

His answer, still in the experimental stage, is a modification of Harden's Ro-Till design that substitutes for the standard subsoiler a shortened shank, at the bottom of which is welded a device made of thin steel that looks a bit like a flattened garden spade. Actually, this humble implement was the prototype. Elkins used a spade in a backyard test plot to make an extremely narrow slit in hardpan just to see if the roots of crop plants could get through it. They could. In fact, Elkins found that roots did not need nearly as much room as the "Y" dug out by a regular subsoiler provides. Accordingly, he fashioned a shortened subsoiler shank that would not itself go through the plow pan but would ride over the top of it and create a tilled planting area, with only the thin steel "slitter" actually penetrating the pan. Thus, instead of a "Y" being created, Elkin's unit makes a "V" with a hairline crack through the hardpan and into the subsoil below the bottom of it. "Once the roots get through that slit and into the loose subsoil, they can go on about their business," Elkins says. "All slit tillage does is put in a macropore." He stresses that the slit has to be narrow so the roots can make good contact with the soil on their way through the pan. Then, after harvest, the roots die in place, plugging the slit with organic matter, which keeps the slit open for future percolation of rainwater, but which, being narrow, maintains the structural integrity of the plow pan. Thus the pan can continue to do its work of protecting the loose subsoil below it from the heavy axle loads pressing down from above.

There is one problem. The slitter is so thin—Elkins says that five thirty-seconds of an inch is the absolute maximum and that he'd prefer it even thinner—that ordinary plow steel is too weak and can wear down rapidly in abrasive or rocky soils. While some prototype slit tillers have been fabricated by Brown Manufacturing in Ozark (the same Brown that produced Harden's first Ro-Till), Elkins believes the device will not be fully viable until a stronger material can be found that is not excessively expensive. There are, however, some offsetting savings to be realized with its use: Elkins says that it takes 40 percent less energy to pull his slit tiller than an ordinary subsoiler. In fact, he says, a "single-bottom" slit tiller can be pulled by a single horse or ox, meaning that it may well become useful in Third World agriculture even before it is widely adopted by commercial farmers in this country.

Despite the drawbacks, Elkins is probably onto something. In fact, Bush Hog is working on its own version of a slit tiller as well as modifying designs for a slimmer subsoil shank.

□   □   □

Meanwhile, fourteen hundred Ro-Till units are already out there, and the effects of their use are being studied carefully by agronomists, who are measuring the impacts of in-row chiseling as performed by the Ro-Till on runoff and yields. In one twelve-year study in Georgia, soybean yields under conventional tillage were 19 bushels per acre but 40 per acre using the in-row chisel. The reason for the great difference in yield is water retention. The runoff on soybean plots under conventional tillage was 33 percent but with the Ro-Till, only 3 percent. Erosion rates went from ten tons per acre per year to zero.

As for that old bottom line, Harden says that with the Ro-Till, you can sometimes double yields, as in the Georgia test, but he adds, more cautiously, that a farmer can consistently get a 10 to 20 percent increase in yields with the Ro-Till at a per-acre cost no

The Ro-Till at work, with uncultivated strips of stubble remaining between the tilled rows. Bush Hog, perhaps alone among conservation-tillage equipment manufacturers, *guarantees* that the Ro-Till will increase yields over the equipment it replaces.

higher, and in some cases lower, than with the equipment it replaces. So confident is Bush Hog of yield improvements with the Ro-Till that as of 1986, they have *guaranteed* a higher yield or "your money back." All a participating farmer has to do is set up side-by-side comparison plots and weigh the resulting yields. Harden believes that the Ro-Till is the only tillage implement to guarantee an increase in production.

Indeed, Albert Trouse, unwontedly throwing a lifetime of academic caution to the winds, says that the results of what he calls "precision applied tillage" via the Ro-Till are "almost too fantastic to believe." It has four great advantages, Trouse says: It "makes fewer trips over the field, reduces wind and water erosion, increases water absorption, and increases yields."

Meanwhile, Harden keeps inventing things. "I'd rather work in the shop than in the field," he confesses to his visitor. And the visitor is aware that he is not being invited into the shop. No doubt some new schemes are hatching with patentable devices to add to an impressive array of patents already awarded—not only for the Ro-Till but for a shielded sprayer and a hydraulic mechanism to operate the "arms" of an articulated spray rig more reliably.

Jerrell Harden is clearly pleased with the success of his Ro-Till, but he is careful to keep his enthusiasm in check. These are pretty good times for the Hardens, the drought of 1986 notwithstanding. But there have been bad times, such as when the dairy herd got Bang's disease, and the carousel dairy barn didn't work, and the whole livestock business had to be abandoned. The dairy buildings remain, empty, a reminder. The Hardens are now "row cropping" full time, growing peanuts and cotton and milo and corn on the same fields as Billy Harden did, five generations and over a hundred years ago. That sort of thing gives a man perspective. Now the sixth generation is already a full partner on the farm—Harden's son, Russell. The succession on the land continues.

Russell's grandfather, J. C. Harden, the historian, speaks for those generations when he says: "The most beautiful sight in the world is a cornfield. From knee high to tasseling time, it's a beautiful sight." It is not always easy to keep faith with this hard, red, gritty land, but the Hardens have done it, and not the least of them has been Jerrell Harden himself.

7

Beyond the Mongongo Tree

*Let us not underestimate the value of a fact; it will
one day flower in a truth.*
—HENRY DAVID THOREAU

It is now time to discuss the environmental implications of conservation tillage.

In human history, there have been two dominant kinds of food economies. One of them might be described as nonmanipulative, a natural economy of benefit to primitive tribes who could collect food from nature without disturbing its balances whatsoever. In his book *People of the Lake*, anthropologist Richard Leakey describes the !Kung people of East Africa as having a food economy of this kind.[1] He calls the !Kung gatherer-hunters rather than hunter-gatherers because the !Kung do not really hunt very much. Nor do they cultivate crops. These aboriginal people live mainly on the nut of the mongongo tree, which grows abundantly on the tops of the dune hills of the Great Rift Valley of Africa, whence, in fact, humans arose as a species distinct from other primates over three million years ago. The average !Kung eats three hundred mongongo nuts a day, a diet that contains the same amount of calories as 2½ pounds of rice and as much protein as a fourteen-ounce steak. Moreover, mongongo nuts are easy to fetch. You just pick them up. There is no danger of overharvesting, for the greater part of the production of

---

[1] The curious spelling of !Kung is to indicate a clicking sound made in the back of the throat, followed by a low musical "ooong."

each tree rots uncollected beneath it. "Why should we plant," said a !Kung to one of Leakey's colleagues, "when there are so many mongongo nuts in the world?"

The other dominant food economy, unlike that of the !Kung, is as manipulative as all get-out. It is agricultural, and though it may have started simply enough with the scattering of seed, it gave way in time to the plow. A plow-based food economy involves perturbing the soil in order to replace native plant associations with monocultures, usually annuals that would have higher food value or greater convenience of harvesting and use or both. As a result, great cities have risen, which is what happens when you can bring food to the people rather than having to bring people to the food. "Where tillage begins," the great orator Daniel Webster observed in an explanation of the origins of civilization, "other arts follow."

In ecological terms, the difference between wandering over to a mongongo tree and picking up lunch and pulling a twelve-bottom plow through a former prairie with a three-hundred-horsepower diesel tractor is immense. One wonders if there is not a middle way. And this may very well be the promise of conservation tillage— finding the middle way.

Ecologically, the difference between nonmanipulative food economies, such as that of the primitive !Kung people in East Africa, and pulling a twelve-bottom plow through a former prairie with a three-hundred-horsepower tractor, as shown here, is immense. The promise of conservation tillage is to help find a middle way.

In *Plowman's Folly,* Edward Faulkner offered a basic theory for a new ecology for agriculture. "We already know—by incontrovertible example," he wrote, "that wherever man does not interfere crops grow spontaneously. It follows of necessity that if man duplicates in his farming the soil conditions which in nature produce such perfect results, he will be able to grow similarly perfect crops on cultivated land." Thus, the ecological principle of conservation tillage would be to capitalize on, rather than eliminate, the natural properties of the soil, which, if "conserved," cán be beneficial in growing crops: structural integrity, porosity, tilth, fertility, and resistance to infestations of pests and diseases.

We know that the environmental impacts of plow-based agriculture have ranged from very high to disastrous—from the loss of the Cedars of Lebanon to the American Dust Bowl. The question is, and it must be asked, Is the new agro-ecology promised by conservation tillage actually a middle way, a partial return to the ecologically benign realms of the nonmanipulative !Kung, or does it simply substitute one kind of adverse environmental impact for another, continuing—maybe even increasing—the serious environmental "externalities" of modern-day commercial farming? At the moment, this question is something of an issue, with the agriculturists, such as Ernest Behn, arguing that conservation tillage significantly reduces environmental impact, and others, such as Maureen Hinkle of the National Audubon Society (whom we will meet later), wondering if conservation tillage, with its use of an arsenal of pesticides, might not produce more pollution than it eliminates.

□   □   □

In the simplest possible terms, current conservation tillage is described as a practice that reduces erosion and agricultural runoff by leaving residues on the surface of the ground. The way this is accomplished, without inviting in a forest of weeds, is by substituting the use of herbicides for the mechanical cultivation needed to get rid of the weeds. Nobody much likes a herbicide, except perhaps those who make them or sell them. For farmers, they are expensive—in fact, herbicides are the most expensive single "input" product farmers must buy—and in hard times they are among the first of the inputs to go when costs have to be cut. Those who believe in more organic ways to grow crops don't like herbicides either because they are, for the most part, about as synthetic as one could imagine, with long, jaw-breaking chemical names (for example, diclofop - [(+)2-[4-(2,4-dichlorophenoxy)phyoxy]propanoic acid], which is a version of 2,4-D). And environmentalists don't like them because they fear they may have unforeseen effects on wildlife or even public health. Paraquat, for example, is on the EPA's list of

The key environmental issues relating to conservation tillage are its implications for the problem of non-point-source pollution from the use of agricultural chemicals and its potential to eliminate soil erosion. It is sometimes thought that conservation tillage is a tradeoff—more chemicals, less erosion. But it's a lot more complicated than that.

pesticide compounds that are harmful to humans in large doses and can pose "an immediate threat to life and health" in the event of an accident at a chemical plant.

And yet by and large, most farmers, most agricultural experts in the universities, and virtually everyone at the U.S. Department of Agriculture believe that the trade-off of herbicides for erosion and agricultural runoff is a good one in terms of environmental quality. They ask whether a small risk associated with increased herbicide use (and to a lesser extent, insecticide and fungicide use) isn't after all justified by the promise that we can, once and for all, finally stop runoff and erosion.

There are two issues here. The first is so-called non-point-source pollution from agriculture. This is the pollution of surface water by, primarily, manure and chemical fertilizer applied to fields. When the rains come and the water runs off, it carries these nutrients with it; phosphorus is the main culprit, although not the only one. Since the pollutants come not from a single point as in a municipal sewage outfall, but from diverse areas as in an agricultural district, it is called a "non-point" source, somewhat nondescriptively. But the effect is the same as from a "point source": the pollution of streams, lakes, reservoirs, rivers, bays, and sounds.

The other issue is erosion—the actual soil material that is washed or blown from agricultural land. Erosion, according to a study by Edwin H. Clark, a water-resources expert at the Conservation Foundation in Washington, D.C., damages the environment in manifold ways. Its impacts are not limited to farms and farmers. Sediment destroys fish-spawning areas in lakes, streams, and estuaries and reduces food sources for all aquatic wildlife, from silt as well as from chemical contaminants. According to the National Fisheries Survey, says Clark, agricultural erosion chronically affects fish life in 30 percent of the nation's waters. In addition, erosion and chemical runoff affect recreational pursuits. Not only fishing but boating, swimming, and hunting are affected to the point of actual economic loss. Erosion damages water-storage facilities, delays commercial shipping (by damaging ships' engines), and requires the dredging of channels. Further, runoff of silt requires the digging out of roadside drainage ditches and increases the cost of pumping water. There does not seem to be any end to it. In total, says Clark, cropland erosion costs us $2.2 billion per year. Clearly, any reduction in erosion, as through conservation tillage, would pay enormous economic as well as environmental benefits to the general public.

As for the effects of conservation tillage on non-point-source pollution, this issue has been the subject of a major study funded by the Environmental Protection Agency. The object of the study was to see if the phosphorus dumped, courtesy of agricultural runoff, into Lake Erie could be significantly reduced by adopting conservation-tillage practices in the lake's drainage basin. As many will remember, Lake Erie was "killed" during the 1960s and 1970s by an algae bloom created by overenrichment of the lake's water from agricultural phosphorus. The effect was to deprive fish of oxygen, creating a kind of North American Dead Sea. An international compact with Canada followed to reduce this severe environmental problem.

The EPA study found that if the amount of cropland in conservation tillage could be increased from the current 25 percent to 47 percent, the United States could, by this means alone, meet the phosphorus-level reduction requirements of the U.S.-Canadian compact—the International Agreement on Great Lakes Water Quality. Moreover, according to the Conservation Tillage Information Center's newsletter, the EPA study showed that the environmental trade-off in producing this desirable effect on Lake Erie would be a good one. The study concluded that phosphorus delivery from runoff could be reduced by a staggering two pounds per acre with "only slightly higher levels of herbicide . . . required to control weed growth in corn and soybean fields," compared with conventional tillage. The CTIC noted that this finding should "dispel fears" of environmentalist opponents to conservation tillage, who were saying that the improvement in erosion and runoff would be outweighed by the chance that some herbicide compounds may get into groundwater supplies.

Maybe not. Maureen Hinkle for one, a policy analyst and lobbyist who works in the Washington, D.C. office of the National Audubon Society, has been slow to accept the case that conservation tillage's net environmental impact is necessarily beneficial. Granting that erosion could indeed be abated, she became concerned, beginning in the early 1980s, about reports that conservation tillage, with its putative dependence on pesticides, could substantially increase the "pesticide load" that has taken a heavy toll on wildlife since World War II.

This was and is a reasonable concern, for the residue on the surface might slow runoff but not necessarily stop it from getting into streams, lakes, reservoirs, or vulnerable bays. While residue on the surface tends to reduce the runoff of silt, it does not reduce the runoff of chemicals to the same degree, since many are water soluble. Moreover, pesticides *not* running off would be sinking into the ground—potentially into groundwater supplies, especially in

Rachel Carson, in her book *Silent Spring*, brought public attention to the problem of agricultural pollution, especially from pesticides. One of the chapters in that book deals with 2,4-D, a herbicide still used in great quantity in conservation tillage. Environmental organizations have continued to be concerned about pesticides and have been effective in banning the use of the most toxic of them. The new chemicals now being produced are not nearly so dangerous as those Ms. Carson wrote about.

sandy soils, where there are fewer clay particles to "bind" the chemicals into the soil.

Why *wouldn't* the Audubon Society want to look into this? On the matter of pesticides, their track record has been impressive. As early as 1945, ornithologist and bird-guide author Richard H. Pough, then a staff member of the Audubon Society, was the first to warn (in a story picked up by the *New Yorker* and others) of the perils of DDT. Later, Rachel Carson raised the alarm in no uncertain terms in her 1962 bestseller, *Silent Spring*. Though some "hard pesticides," especially those like DDT, which tend to remain in the food chain, were banned, others that she warned against were not—including 2,4-D, a mainstay contact herbicide in the conservation-tillage arsenal. Is it possible, wondered members of the Audubon Society and other environmentalist groups, that conservation tillage might bring about a second "silent spring" with a whole new array of pesticides whose volume of use would be encouraged by no-till or associated conservation-tillage practices? New insecticides had been developed for corn borers harbored in the stubble of "continuous corn" in the Midwest; new fungicides had been made to deal with the soilborne pathogens given a new lease on life in the no-tilled wheat fields of the High Plains and the Palouse; new herbicides had been formulated to deal with the weeds that can grow so suddenly and rankly in the warm, red, humid fields of the southeastern Coastal Plain.

Because it is Maureen Hinkle's job to investigate such matters in behalf of the 400,000 members of the National Audubon Society, she took it upon herself to look into the growing literature concerning these new compounds. Here is a small sampling of some of her findings, which she gleaned from hundreds of scientific-journal articles:

¶ Insects, rodents, nematodes, fungi, and other pests tend to increase with conservation tillage, requiring as much as a 30 percent increase in the use of pesticides over conventional tillage.

¶ Atrazine, a herbicide that is applied to 50 percent of corn acreage in the North Central states, injures small grains, soybeans, or alfalfa planted the following year, locking farmers into continuous-corn sequences.

¶ Repeated use of a single herbicide can induce rapid development of resistant weeds. Sicklepod, once a minor weed in South Carolina, became a major weed by 1981 after several years in no-till.

¶ Extremely high levels of toxaphene (an insecticide used on no-till corn for armyworms, cutworms, and borers, among others) found in the Yazoo River in Mississippi led the EPA to conclude that people eating fish from the Mississippi Delta were at high risk of developing cancer.

¶ Complete metabolic-breakdown pathways are known for only 3 or 4 of some 150 chemicals used in herbicides. Metabolic degradation of the remaining compounds is only partly known.

These findings, somewhat condensed and paraphrased, are taken from a heavily footnoted article Ms. Hinkle wrote for the *Journal of Soil and Water Conservation*. And, as might be expected, some quite colorful reactions showed up in her mailbox following publication. While in a few cases her facts were questioned, for the most part her motives were at issue. One writer wanted to know why she was "bad-mouthing no-till" when, compared with conventional tillage, it conferred so many benefits to both farmers and consumers. Would she have us return to the evils of the moldboard plow? Actually not. What most environmentalists would prefer is that the extremely erosion-prone land be taken out of agriculture altogether, not held hostage to conservation tillage with its dubious dependence on chemicals.

The National Audubon Society is not, of course, the only environmental organization to wonder about the down-side risks of conservation tillage. Jack Doyle of the Environmental Policy Institute testified before a 1985 congressional hearing that although the herbicides currently used are not thought to be as toxic to humans as the persistent chlorinated hydrocarbons of the past (such as DDT), "not much is known about the long-term effects of herbicides in the environment." Doyle says that Lasso, Roundup, and paraquat have shown up in groundwater studies in half a dozen states. "Anywhere they look for it, they seem to find it," he says. He cited a 1984 study by the prestigious National Academy of Sciences, which concluded that current toxicological data were simply insufficient for them to permit even a partial assessment of the effects on human health of 66 percent of the ingredients of pesticides (including herbicides) now in use.

According to Donna Fletcher, who headed a task force at the U.S. Environmental Protection Agency to look into the appearance of the new herbicides in groundwater, "the jury is still out" on whether the new compounds are as safe as they ought to be. Ms. Fletcher makes another point: it's not just a matter of how many pounds of pesticides are used on a given piece of land, but the staggering number of reactive combinations these chemicals make with the soil and with each other in terms of toxicity for humans.

The environmentalists in and out of government seem to agree on two things: the new herbicides don't disappear as readily as their makers and users often suggest that they do, and the issue of human toxicity is still very much a puzzle.

Environmentalists in and out of government are concerned about the unknown ways in which pesticides, including the herbicides used in conservation tillage, react with one another and with the soil in various situations. Here, a government scientist is studying the breakdown of pesticides in wastewater in a field test. According to one expert, "herbicides are persisting in groundwater year-round."

The EPA does not believe that most of the herbicides are necessarily dangerous to the general public. Only a few herbicides are on the EPA's "danger list" of agricultural chemicals. Even so, the agency still wants "to make sure that the solution to one problem does not cause another," according to an EPA policy-staff member. The problem, from the EPA's point of view, is not only soil erosion, but water pollution of two kinds—non-point-source pollution and groundwater pollution from the seepage of pesticides into subsurface water-bearing strata. EPA staff member Timothy Amsden told members of the Institute for Alternative Agriculture that the agency would begin in the spring of 1986 to "conduct rigorous surveys to see the extent of the problem." He told the group that the EPA had "called in leachability information" for the pesticides on the danger list "and will put that together with toxicity and persistence information to determine what needs to be done."

To assist in evaluating *new* pesticide compounds for efficacy and possible toxicity, the EPA and the USDA's Agricultural Research Service have developed an elaborate computer model that can handle all the variables involved in the biochemistry of the compounds once they hit the soil. The model will, says William F. Spenser of the ARS, be able to trace "all possible routes a pesticide might take, just as though it were released into the environment."

In a roundup paper on agricultural pesticides and water quality for the National Academy of Sciences, George R. Hallberg of the Iowa Geological Survey cites the occurrence of nine herbicides (although not paraquat) in groundwater sample studies done in twenty-three different states. The pesticide-state groundwater matchup is as follows.

Alachlor*	Maryland, Iowa, Nebraska, Pennsylvania
Aldicarb	Arkansas, Arizona, California, Florida, Massachusetts, Maine, North Carolina, New Jersey, New York, Oregon, Rhode Island, Texas, Virginia, Washington, Wisconsin
Atrazine*	Maryland, Iowa, Nebraska, Wisconsin
Bromcil*	Florida
Carbofuran	Maryland, New York, Wisconsin
Cyanazine*	Iowa, Pennsylvania
DBCP	Arizona, California, Hawaii, Maryland, South Carolina
DCPA*	New York
1,2-Dichloropropane	California, Maryland, New York, Washington
Dinoseb*	New York
Dyfonate	Iowa
EDB	Arizona, California, Connecticut, Florida, Georgia, Massachusetts, South Carolina, Washington

Metolachlor*	Iowa, Pennsylvania
Metribuzin*	Iowa
Oxamyl	New York, Rhode Island
Simazine*	Maryland, Pennsylvania, California
1,2,3,-Trichloropropane	California, Hawaii

*Herbicides. The others are insecticides and nematicides.

According to Hallberg, the groundwater studies in his home state of Iowa show that "many of the most commonly used pesticides, particularly herbicides, are leaching into groundwater in a variety of environments. The findings indicate that several of the herbicides are persisting in groundwater year-round."

There is worse news. Agricultural chemical companies, eager to continue the conservation-tillage bonanza, are going in a brand-new direction—toward genetically engineered hybrids created especially to be resistant to herbicides. Thus far, a bioengineered soybean that suffers no ill effects from Roundup—a quite popular systemic herbicide—is near, and a tobacco cultivar with the same resistance is ready for field testing. There are vast economic stakes in this concept, says Jack Doyle of the Environmental Policy Institute, and for this reason giant chemical corporations, including Monsanto, American Cyanamid, and DuPont, are moving into the seed business, taking over what before were largely family-owned companies. The result, environmentalists fear, is the perpetuation of wholesale drenching of fields with herbicides—at whatever the cost in dollars and environmental risk—simply because there are crop-plant cultivars available that are immune to them.

□   □   □

Despite the foregoing, all of which understandably alarms environmentalists quite seriously, the urgent and historically perplexing agricultural problem remains: what to do about erosion. Since the 1930s, the U.S. government has spent tens of billions of dollars to stem the ravages of wind and water on the land, and yet erosion continues apace. Between 1977 and 1982, the most recent period studied by the USDA's National Resources Inventory, 3 billion tons of soil were annually lost to erosion: 1.2 billion tons to wind and 1.8 billion tons to water. In all, according to USDA officials, this amount of topsoil could cover one million acres to a depth of one foot. And so the apologists for conservation tillage may be forgiven for emphasizing, again and again, the astonishing statistic that up to 90 percent of soil erosion can be reduced where such tillage is practiced.

Unfortunately, the areas where conservation tillage is most popular among farmers and the quite limited areas where erosion rates

This photograph was taken in 1936 but could have been taken any time in the past fifty years. The United States annually loses 1.2 billion tons of soil to wind and 1.8 billion to water. Because conservation tillage can prevent such erosion, many do not believe that its use should be held back, despite the problems of pesticides associated with its use.

are extremely high do not perfectly match. Pierre Crosson, an economist at Resources for the Future, a Washington, D.C. think tank, asserts that "much conservation tillage is on non-erosive land and little of it is on the most erosive land." While this analysis, originally developed by the U.S. Office of Technology Assessment, reflects a statistical disjunction rather than any inherent difficulty in the adoption of some form of conservation tillage on erosion-prone land, some environmentalists have latched onto it to show that there is no real trade-off and therefore soil conservation via conservation tillage should not be used as an excuse for increasing the pesticide load. Perhaps they are right, although not for this reason. It is well to remember that neither Faulkner nor any modern practitioner of conservation tillage suggests that the primary impetus for "plow-less farming" is to reduce external impacts. It is undertaken as a new agro-ecology that has multiple agronomic benefits. To the extent that there are public ones, they are a bonus.

Nevertheless, the fine points of public environmental benefit, however arcane they may be to farmers, are the petard the industrial-agribusiness-oriented establishment insists on using to hoist the environmental movement. The only alternative to the destructive moldboard plow, the agribusinessmen argue, is a herbicide-based tillage system that leaves residues on the surface of the soil. To be against herbicides, therefore, is to be for a return to clear tillage and therefore to favor continued erosion and pollution from agricultural runoff. Stalemate.

□   □   □

The trouble with this trade-off argument is that it assumes that the ecology of conservation tillage is static when in fact, it is quite dynamic. It assumes that *intensive* herbicide use is now, and will continue to be, the sine qua non of this new agricultural ecology. In this regard, environmentalists and chemical companies are in unwonted, as well as inadvertent, agreement. The only difference between them is that the chemical companies believe this is wonderful ("No-till farming is our number one priority," says one Chevron executive), and the environmentalists think it's awful.

To determine the fact of the matter, one must begin with the farmers themselves. David Schertz of the USDA's Soil Conservation Service talks with farmers all over the country about the on-the-ground practice of conservation tillage. His findings lead him to say flatly that the notion that conservation tillage increases pesticide use is a "fallacy." There is, to some degree, an increase in herbicide use in the early years of conservation tillage, especially no-till, but most farmers find that year by year, the weed problem gets less and less. This is partly because weed seeds are no longer turned up by plowing. But improvements in technique, too, can reduce herbicide use. For example, Jerrell Harden's Ro-Till system is able to incorporate herbicides efficiently into the soil, not possible with earlier conservation-tillage implements. Incorporation reduces the herbicide load by "banding" the chemicals so that they work only on the target weeds rather than relying on an allover spray. The incorporation of the band also reduces the possibility of chemical runoff since the herbicides are not on the surface. Ridge tillage, as practiced by Ernest Behn and the Eppleys, also reduces herbicide use substantially. This obtains because of a combination of mechanical tillage on the ridges and precise herbicide placement between them at planting and because at midseason another pass is made to rebuild the ridges, which can clear the row middles of weeds completely without herbicides. It is quite likely, in fact, that as ridge-tillage techniques develop, this method of cultivation can consistently use less herbicides, insecticides and chemicals of any kind, including fertilizers, than the tillage practices it is so rapidly replacing in the Corn Belt.

As for new postemergent herbicides, formulated to take out a narrow spectrum of weeds after the crop plants have already "emerged," new techniques make it possible to use spot applications on a remedial basis rather than relying entirely on a wholesale preventive regimen. Jerrell Harden's shielded sprayer serves this purpose. Another device used in a postemergent spraying system is an electric eye. When the eye sees a weed in a row, the sprayer gives it a shot, but otherwise it is turned off. Yet another technique has been developed involving the very efficient systemic herbicides: a

New application techniques for herbicides in conservation tillage, like this rope
wick device, can do much to reduce the total amount of chemicals used. There
is virtually no possibility of runoff with this system.

small amount of herbicide is applied to a weed's leaves, after which
it begins "translocating" throughout the vascular system, killing
the plant to the roots. Obviously, such herbicides cannot be sprayed
in a field with a standing crop lest drift or bad aim kill off the
wrong plants. Traditionally, the systemics have had to be used on
a preemergent basis, by spraying a whole field prior to planting.
But now, an applicator has been developed that dangles a wick from
a boom that brushes the tops of the weeds between (or even above)
the rows. There is virtually no possibility of runoff with this system.
Moreover, only a fraction of the amount of herbicide is needed compared with an edge-to-edge dousing of a field with a weed killer.

Such advances in herbicide applications are not yet common.
The vast majority of conservation-tillage farmers simply trundle
out the tank and spray the fields. But the new techniques are coming, and some agricultural experts believe that environmentalists
should learn about them. Says Washington State's Robert Papendick, who sometimes lectures to environmental groups: "Anybody
in resource conservation should get behind this movement. I think
we've got to squelch this idea that no-till is necessarily tied to
increased use of pesticides. That may be the fact today. But it need
not be the fact tomorrow. No-till is not going to make it in the future
if its feasibility continues to depend on the increased use of pesticides."

Papendick is working on new rotations involving new cover crops to control pests in conservation tillage. "We're going to have to pay a lot more attention to crop-rotation systems with conservation tillage than we have with conventional tillage," he says.

In search of new ideas, Papendick journeyed to Australia, where a biennial legume called "black medic" is used in rotation with sheep grazing and wheat. Papendick believes that black medic might well be alternated with wheat as pasture or green manure in dryland farming in the United States. "What we're attempting to do," he says, "is develop a system that is highly competitive with weeds so that we can minimize herbicides—use them only when there is a runaway epidemic.

"Weeds are a symptom of a problem," Papendick continues. "They mean that something has gone wrong with the system, that what you are doing is more favorable to the weed than to a crop. You should be able to figure out a way to make the system more favorable to the crop."

Papendick has studied Mort Swanson's paired-row wheat-planting system described in Chapter 4 and believes that it can greatly reduce chemical inputs. In addition, he is investigating intercropping the area between wheat rows with legumes. This way, the legumes compete with the weeds and provide nitrogen for the next season's crop. "The whole idea," says Papendick, "is to cut down on herbicides and fertilizer."

A colleague of Papendick's, weed scientist Alex Ogg, makes some telling points in this regard. Ogg, born and brought up on a small Wyoming farm, is no apologist for herbicides, as perhaps some of his professional brethren are likely to be. "The public," says Ogg, "are not going to stand for more and more pesticides, and anybody who says they will is just blind to the facts."

Ogg believes that the best of the herbicide manufacturers have gotten this message pretty well. He is particularly keen about a new generation of herbicides based on sulfonylurea—a new "line of chemistry" developed by DuPont. The application rates for these compounds are measured by the gram or in a graduated cylinder. "Some of the new products," says Ogg, "use only one-sixteenth of an ounce per acre. Half a pound—two hundred grams—can handle a hundred-acre field." While the molecules of these new compounds are "very active," Ogg says that by using a fraction of an ounce per acre rather than the old pints and quarts of traditional herbicides, the new compounds "should substantially reduce the potential for environmental contamination."

Ogg is also pleased that DuPont has chosen to make the cost of the sulfonylureas quite moderate. In fact, they are priced competitively with 2,4-D, one of the least expensive of the contact herbicides. But the sulfonylureas are not limited to a narrow range of use—they can, says Ogg, be formulated in preemergent and post-emergent forms and can be modified to be "long or short" residuals. Moreover, the new chemistry permits "designer" formulations for certain crops or certain weeds. This too reduces the risk of environmental contamination, since they would be used more efficiently than broad-spectrum compounds. Moreover, greater specialization in the formulations helps guard against the emergence of weeds that are resistant to herbicides.

Another front-edge advance in reducing the environmental impacts from herbicide use is the concept of coating herbicides directly on seeds, which could, says Ogg, "reduce the total amount of pesticide per acre by a large margin." Moreover, Ogg is watching the development of biological herbicides with interest. One of these is a fungus, found naturally in southern agricultural areas, that when "cultured" and sprayed can control the vicious sicklepod, the weed that Maureen Hinkle's article reported had become such a major pest in no-till fields in South Carolina.

It would appear, given the foregoing, that the "hard-herbicide" era of conservation tillage, characterized by the use of the highly toxic and persistent paraquat, may be on the way out.[2] Although anyone reading a farm magazine would be likely to conclude that chemical companies have a mortal lock on U.S. agriculture in general and conservation tillage in particular, in fact the grip is being loosened by agricultural-chemical companies themselves. In a curious irony of free-enterprise economics (albeit one that is modified by EPA rules and regulations and a number of energetic environmental groups), the chemical companies may, for the sake of staying competitive in the growing conservation-tillage market, be showing farmers the way toward a reduced pesticide load rather than an increasing one, with safer, more efficient products that are no

---

[2]Paraquat comes in for such a drubbing by environmentalists and even weed scientists like Ogg, that ICI, the giant British chemical manufacturer that invented paraquat in the 1950s and, along with Chevron, distributes it in the United States, has published a booklet called "Paraquat in Perspective." In it, ICI states that para-quat is "an environmentally sound product and does not adversely affect health when used in accordance with label directions." As for its persistence in the soil, the booklet offers this explanation: "Paraquat reaching the ground is bound to clay particles in the soil within minutes and is biologically inactivated; that is, it no longer has any activity on plants, the soil, or other organisms. In the longer term, it is slowly decomposed in the soil. It is possible that paraquat that is bound to the soil could be washing into streams or ponds. But it would be biologically inert, tightly bound to clay particles or bottom mud." The booklet is available from the Director of Public Affairs, ICI Americas Inc., Wilmington, DE 19897.

The greatest impetus for more ecologically benign pesticides has been, according to one expert, "the pressure exerted on the chemical companies by environmentalists over the years," working through the U.S. Environmental Protection Agency. The result is that massive-scale farm operations like this one can use pesticides that are measured by the gram per acre rather than by the tankful.

more expensive than the environmentally damaging compounds they replace.

According to Ken Cook, agricultural policy expert at the Conservation Foundation, a Washington, D.C. environmental-research organization, the greatest impetus for this change has been "the pressure exerted on the chemical companies by environmentalists over the years, primarily via the provisions of FIFRA [the Federal Insecticide, Fungicide, and Rodenticide Act] and the Environmental Protection Agency's pesticide registration program." As Cook explains it: "Big chemical companies need new pesticide products all the time because patents run out after seventeen years and then the me-too companies come along and eat their profits (which are enormous). Because of environmentalists like Maureen Hinkle, as these new products come along, they now have to go through a twenty-five-million-dollar-plus research-and-testing program designed to make them safe for humans and the environment. That process scarcely existed fifteen years ago, which is why we have so very little information on the older chemicals." The result is that the chemical companies, unwilling to make such an extraordinary investment in herbicides and other compounds that may not pass environmental muster, are now more likely to create products that are a great deal more ecologically benign than those they are replacing, even as they strive for increased efficacy. This is, perhaps, a good illustration of the way a free society is supposed to interact with a market economy.

Says Alex Ogg: "Pesticides are not inherently bad, if used properly. Toothpaste and deodorants are pesticides, after all, and we use them every day. But we have had too much dependency on them in agriculture. One of the things that herbicide users have to face is that you can't just keep spraying to solve weed problems. Sooner or later, you have to get back to doing a good job of farming. Healthy crops are always the best answer to weeds."

□   □   □

Organic farmers and their supporters would agree that agriculture has too much dependency on herbicides, as Ogg says, but would disagree with the thought that they are not inherently bad. Herbicides, say experts in organic agriculture, can alter crop physiology, tending to weaken plants so that they are less able to resist infestations of insects or soilborne pathogens. Moreover, continued use of herbicides can alter populations of beneficial microorganisms in the soil.

For these reasons and others, organic farmers have tended to stick with conventional tillage, with its reliance on mechanical means of weed control. Thus, most organic-farming people and most conservation-tillage people have agreed that the two techniques must

The Rodale Research Center in Kutztown, Pennsylvania, where Barney Volak is experimenting with allelopathic cover crops that can be used instead of herbicides in conservation tillage.

inevitably take separate paths to the reduction of erosion and agricultural runoff.

There are, however, some who believe in investigating the possibilities of convergence. Among these is Barney Volak of the Rodale Research Center in Kutztown, Pennsylvania. The center is part of Rodale Press, publishers of *Organic Gardening* magazine and national leaders in the organic-farming movement for many years.

Volak is testing various alternatives to herbicides in conservation-tillage regimes: different crop rotations, biological and predator controls, and the use of a modicum of mechanical tillage. A fourth technique being investigated is the use of crop competition to control weeds. This approach includes shading provided by a ground cover; the use of high-quality, vigorous cultivars; and more closely spaced plants in the row, with the rows themselves moved closer together. The tests are just beginning, but a preliminary finding, reports Volak, is the achievement of "competitive yields from corn grown with alfalfa" in ridge-tillage and no-till tests.

Of great interest to organic-farming experts as well as those involved in conservation tillage are the possibilities offered by "allelopathic" cover crops. A major Rodale test is to join conservation-tillage methods with "allelochemicals"—"natural pesticides," as Volak calls them—to control weeds in corn-production systems. "Allelo" is a Greek combining form indicating reciprocity; hence, allelopathic plants are reciprocally pathological to one another.

Other scientists too are finding new opportunities in allelopathy. According to Douglas Worsham of North Carolina State University, as simple a thing as planting a cover crop of annual rye, but leaving it on the surface after it dies rather than plowing it under, produces compounds that can slow the growth of weeds—without inhibiting crops planted into the residue with a seed drill, since they are placed beneath the surface and out of harm's way. James D. Riggleman, a weed scientist from the DuPont Company and 1986 president of the Weed Science Society, suggests that new allelopathic substances might even be extracted from plants or produced synthetically for use as a herbicidal spray or for incorporation into the soil.

Riggleman proposes, in fact, that "new ecological, biological, and nonchemical methods of weed control" be a top priority for all weed scientists. "While biological control of a broad range of weeds is unrealistic at current levels of technology, we have positive examples in the control of single species." Riggleman says that biological control of five weed species is now feasible and that by the year 2000, thirty more might be added to this list. One fascinating idea advanced by Riggleman in his 1986 presidential address to the

weed scientists was that weed-seed germination might be influenced biologically, which could, as he put it, offer "a remarkable
new approach to weed control." Riggleman uses cocklebur as an
example. If cocklebur seeds could "be forced to germinate very
early or very late in the season, or even all at one time, the management of this pest would be greatly simplified." He does warn his
colleagues, "Don't hold your breath" for the development of this
technology, but it serves as a good example of the way conservation
tillage can reduce its dependence on herbicides and begin to con-
verge with the ideas and ideals of organic farmers.

Riggleman is no proponent of organic farming, of course. He
foresees the development of "herbicide-tolerant cultivars from conventional plant breeding and genetic engineering." Some such cultivars, Riggleman says, are already under way: tobacco, tomato,
corn, and soybeans. "We need greater crop tolerance through con-
trolled-release formulations, antidotes, or safeners," he says; but he
adds, however, that the "main issue" in herbicide safety is still and
will remain the care with which the chemicals are applied in commercial agriculture. "Precision, foolproof application remains as
probably one of the most-often-overlooked areas for improvement.
There is strong support for expanded application research in the
USDA and the states, as well as in industry."[3]

□   □   □

Despite the long-term prospect of significantly reducing environmental impacts from agricultural herbicides, a wholly nonchemical means of dealing with weeds in conservation tillage, such
as those proposed by Barney Volak, may seem wildly theoretical to
most agronomists and farmers. The fact is, however, that there are
at least two commercial farmers in the United States who are trying
"organic conservation tillage" with some success.

One of these is Dick Thompson, who runs a three-hundred-acre
farm near Boone, Iowa (Ernie Behn's hometown). Thompson is a
contributing editor to a Rodale magazine called *New Farm*, which
reports on the doings of alternative agriculture under the auspices
of Rodale's nonprofit Regenerative Agriculture Association. "Herbicides have not been used on this farm since 1967," says Thompson,
but he tested out ridge tillage with a Buffalo till planter (such as

---

[3]One agricultural scientist who would agree with this goal wholeheartedly is
Purdue's William C. Moldenhauer, who has a high level of concern for the environment. Moldenhauer believes that much if not most of the pesticides now showing
up in groundwater get there not by leaching, but by carelessness on the part of
applicators, who have been known to flush spray tanks directly into abandoned
wells, thus moving the compounds directly into water-bearing strata.

Dick Thompson, shown here on his Iowa farm, is an organic farmer who is showing how an "organic" form of conservation tillage—actually a variation on ridge tillage—can be practical and profitable.

the Eppleys use) anyway to see if he could adapt it to his kind of farming. "In three soybean growing seasons," he reports, "side-by-side comparisons have shown that conventional tillage, fall and spring, produced more weed pressure in both broadleafs and grasses. On the Buffalo ridge-strip till plant plot it was easier to manage weeds than on the conventional, even without using herbicides."

One principal difference between Thompson's ridge-till system and those that apply herbicides is the use of a rotary hoe. "The rotary hoe has always been our first choice for postemergent weed control," says Thompson, "since it can be used on a wide range of soils and crops without excessive damage to either. Two rotary hoeings are normally enough." While this involves one more trip across the field during growing season than does ridge tillage using herbicides, compaction should not be increased since, in ridge tillage, the wheels are restricted to the space between the rows and do not run across the growing area.

In addition to the postemergent rotary hoeing, Thompson controls weeds by means of a light harrowing in the very early spring so that before the till planter comes along a few weeks later, early weeds will have "expressed themselves" and can be taken out by the planter's sweeps as well as by a preemergent rotary hoeing. This procedure would substitute for the special "burn-down" herbicide spray application that some ridge tillers do before planting. The planting itself is, for some crops, at a higher density than most farmers use, to provide better crop competition against weeds. The seed costs are higher, but they are offset by the savings on herbicides. In the fall, Thompson plants cover crops with allelopathic properties, including oats and rye: "Both crops control erosion and add valuable organic matter to our soil, and we've found that they chemically suppress weed growth in the corn that follows. The oats winter kill, and the small amount of rye left over is just enough to help control early weeds without clogging up the planter sweep."

In the end, Thompson gets competitive yields at a much lower cost. "One feature that sets ridge-tillage apart," says Thompson in a *New Farm* editorial, "is that it's the only reduced till method with the potential for eliminating groundwater pollution from herbicides and pesticides. . . . What this means is that, while it may not have been officially announced until now, the marriage between agriculture and conservation has already yielded offspring that we can be proud of."

Another farmer working on the convergence of conservation tillage and organic farming is Richard Harter, a California rice farmer who, in the same year as Thompson, "went organic." According to Patrick Madden, an agricultural economist at Pennsylvania State University who has studied the operation firsthand, Harter, while walking through his fields, "became disgusted with the smell of chemical poisons he had been applying. He realized that these poisons were actually destroying life on his farm. . . . He yearned to see the butterflies, to hear songbirds . . . that had originally thrived on his land."

In 1982 Harter, by then fully "organic," began experiments with no-till organic rice, overseeded into a legume, and by the next year began using no-till for his entire "adobe-ground" rice crop. His method, says Madden, "is based on the same general idea that chemically oriented farmers are increasingly using to grow corn interplanted into an established sod crop. But instead of using herbicides to suppress the legume growth, Dick floods the fields."

For "upland" rice, in which fields are not flooded, Harter uses a no-till method involving matting down the residue of annual weeds from the previous year, seeding heavily and quadrupling the amount of fertilizer (chicken manure). The result is a heavier stand of rice that can better compete with weeds. While the productivity of the organic no-till method on the adobe ground was comparable to nonorganic yields in the area—sixty to sixty-five hundredweight per acre—upland yields were at first less than half that. More recently, fine-tuning has increased the harvest. Reports Madden, "Dick has been very pleased with the results of his latest no-till rice effort. During a visit to his farm in September, 1985, I found that his upland no-till organic rice stand was very heavy, with very few weeds. A follow-up conversation indicated his yield was thirty-seven hundredweight."

□   □   □

Despite the work of pioneers like Thompson and Harter, it is highly doubtful that commercial conservation tillage will ever be practiced widely on an entirely herbicide-free, organic basis, given the greater costs, greater manpower requirements, and risks of lower yields. At the same time, these experiments in organic conservation tillage suggest that the "truism" that the basic idea is simply to substitute herbicides for mechanical tillage is not really true at all. Nor is it a sound basis on which to evaluate future, or even present, environmental impacts of conservation tillage.

Nor is a corollary truism true either—that conservation tillage can be defined entirely in terms of the amount of residue left on the surface of the ground (even though it has been described this way in earlier parts of this book). This is a bit like describing the complexities of a building—its foundation, its many rooms, its heating and cooling and plumbing—in terms of the thickness of its shingles. Conservation tillage must be understood in terms of what one does *underneath* the ground as well as what one leaves on top of it. Conservation tillage in its fully realized forms—no-till, ridge-till, and strip-till, for example—leaves both the top and the bottom relatively undisturbed compared with plow agriculture.[4] And where the ground is not massively disturbed, the dynamics of the soil, with gentle nudges by the agriculturist, can be made to produce crops reliably for as long as the agriculturist wishes. It is this quality that makes the technique "revolutionary," not the presence or absence of herbicides nor even the measurement of the amount of residues. Residues are essential to the ecology of conservation tillage but do not, in themselves, define its complex dynamic.

It is this dynamic—the dynamic of soil being allowed to do its work naturally in the growing of crops—that can make conservation tillage as ecologically different from plow farming as plow farming is from the food gathering of the !Kung, with their wonderful mongongo trees. With herbicides measured by the gram, new devices for applying chemicals, new biological weed controls, new rotations with allelopathic cover crops, and even *organic* no-till, a new phase in the ecology of conservation tillage is about to

---

[4]Chisel plowing and related techniques are usually included as forms of conservation tillage because they leave at least 30 percent of residue on the surface. Ecologically, however, some chisel plowing (depending on depth) may be more akin to plow-based agriculture.

What makes conservation tillage "revolutionary" is simply letting the soil work naturally at growing crops. It is this that makes conservation tillage as ecologically different from plow farming as plow farming is from simple food gathering in primitive societies.

begin. The directions suggested by Robert Papendick, Alex Ogg, James Riggleman, Barney Volak, Richard Thompson, and Richard Harter appear, on balance, to be quite promising from the standpoint of environmental quality.

Edward Faulkner defined plowless farming as a "new agriculture which is in reality very old." Civilization has brought us well beyond the mongongo tree; but with the emerging ecological concepts of conservation tillage, we may yet recapture some of the environmental gentleness we have lost during the millennia we have been using the plow.

8

Old Farms, New Lives

Let us now praise famous men
—ECCLESIASTICUS 44:1

"Vermont," writes Charles Morrissey, a former state historian and author of the engaging *Vermont: A History* (one of the "States and the Nation" books commissioned for the U.S. Bicentennial), is "a reminder of the American past, a remnant from the agrarian culture which once we were." And so it is—a good place to go to be reminded how very unstable and transitory the practice of agriculture has been in the brief history of this nation.

Today in Vermont, the old hill farms are almost gone, grown to brush as the land passes out of active agricultural ownership and use and becomes a weekend place, a piece of real estate held for speculation and, in a good many cases, developed for nonagricultural uses: Hill Farm Condos. In the process, according to George Dunsmore, a farmer and at one time Vermont's commissioner of agriculture, "The forest cover in Vermont has gone from twenty-eighty to eighty-twenty." By this he means that those earliest settlers who came up the Connecticut River valley had managed, within a century or less, to clear four-fifths of the land of trees and rocks so that they could grow crops and pasture animals. Early on, Vermont was the chief producer of wheat—in Colonial times it was the breadbasket of the Colonies (though Vermont was not one of them). Later on, Vermonters specialized in sheep, a natural for the steep, wet pastureland. But they lost that business too. The Erie Canal, and later the railroads, opened up the western lands, and so Ver-

Vermonter John Deere left the hill farms of New England to establish his plow factory on the Illinois prairie. Agriculture was on the move, and it has stayed on the move ever since.

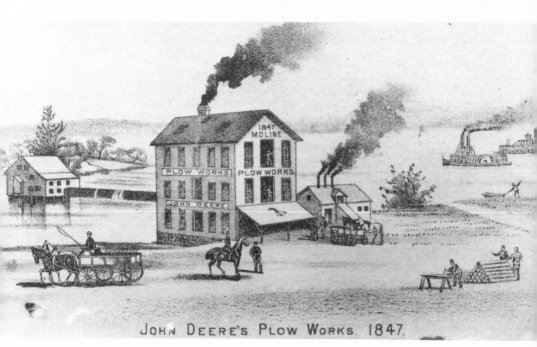

JOHN DEERE'S PLOW WORKS. 1847.

mont went in for cows. At first, they had a corner on cheese, but stiff competition came from Wisconsin and elsewhere so that now they specialize in milk, which, because it is perishable, cannot be so easily overrun by midwestern or western producers.

The hill farms started going under in Vermont with the mass migration after the Civil War to the flat land that was discovered in Illinois and Iowa. And then the forest cover began to come back; now it once again blankets the landscape, or at least 80 percent of it, and the old houses and barns are rotten, with only the cellar holes remaining, and the stone fences that once wound through the pastures are concealed by the second growth of brush and hardwood saplings.

"An aura of tragedy seems to hover over the Vermont landscape," writes Charles Morrissey. "You sense it on seeing a crumbling mill-race . . . in the beeches and balsam grown up in what was once somebody's pasture . . . in the space left in the family burial plot for the youngster who went west and never came home."

The sad, abandoned farms of Vermont. "One of the most moving trips you can take in Vermont," says Stephen Kerr, a former associate of Commissioner Dunsmore, "is to take Route 125 up to the top of the Green Mountains to Ripton and just look at the remains of the farm fields that were up there. You get out on the escarpment and look across the valley and you can almost picture what it was once like. What you had was hundreds of farms on that shallow but very loose and tillable soil. Those places were farmed for about fifty years, until they literally exhausted the land. Back then we didn't have lime; we didn't have chemical fertilizers. And a lot of people actually starved to death. Look at the gravestones. Farmers died at the age of thirty, kids at six months."

And then John Deere came along. As recounted in Chapter 3 of this book, there is a theory one hears in discussions of agricultural history that John Deere, a Vermonter from Rutland, was personally responsible for changing Vermont—and all of New England for that matter—and that moving west, he permanently modified the course of history of the young nation by mass-producing the steel mold-board plow. Born in 1804, Deere apprenticed as a blacksmith at the age of seventeen and later set up his own smithy in Vermont. But in 1837, he followed those farmers whose land had run out in the hillsides around Rutland. It was in Illinois that he first produced the famous steel plow that, with a stout team, could turn the heavy bottomland soils along midwestern rivers and slice through the matted sod of the prairie, opening new territories that had formerly been unavailable for "land farming," as they call it still in Vermont.

And so, beginning in rocky New England hillsides and the piedmont soils of the Appalachians, agriculture marched across the face of America, used up the land, moved on, and used up more, until even places utterly inhospitable to agriculture had to be pressed into service at great public cost—for irrigation dams, diversions, and pipelines. Homesteading, Resettlement, and Reclamation were federal policies that insured that agriculture kept on the move. To the European colonists, the New World was to provide land aplenty for agricultural enterprise, there for the taking—land that would in a thousand years still not be brought fully into use, or so thought Thomas Jefferson. Instead, all of it was not only used, but a significant fraction used up, in three hundred.

Although most Americans insist on thinking of agriculture as a solid and stable way of life, it is anything but. The old farms are rare; they are the exception rather than the rule. And we celebrate the exceptions, like Jerrell Harden's five-generation land in Alabama, perhaps to avoid confronting the rule, which is found in the graveyards of Vermont, the desertified farms of the Southwest, and the abandoned farmsteads everywhere—barn roofs caved in; roads gone to weed and ruts; farmhouse porches rotten and askew; upstairs window glass blown out of dark gray, weathered frames; a lonely,

Abandoned farms have been a hallmark of U.S. agriculture, an economic pursuit and way of life that should be stable rather than tenuous. This former farm is in Adams County, Ohio.

dusty, tattered curtain stirring faintly as if in a weak and final wave of farewell.

Like other industries—coal and iron, for example—that depend on exploiting a resource and then moving on, the enterprise of farming is about as stable as a chip in a Vermont millrace. What a contrast this unrooted, *extensive* form of agriculture is to the *intensive* European agriculture it replaced. René Dubos, in his book *A God Within*, describes the place of his growing up, the Île de France country north of Paris, near Picardy and Normandy, as a land "that has remained very fertile, even though much of it has been in continuous use for more than two thousand years."

Dubos quotes the French poet Charles Péguy:

> *Deux mille ans de labeur ont fait de cette terre*
> *un reservoir san fin pour les âges nouveaux.*
> (Two thousand years of human labor have made of this land an inexhaustible source of wealth for the times to come.)

□   □   □

What does all this have to do with conservation tillage? Perhaps not much in the short term, but in the long term, the tillage revolution just may be the first step toward an American agriculture that is not only prosperous, as it has been at times in the past, but stable too, which it actually has never been. In fact, prosperity and stability have seemed almost mutually exclusive throughout the history of American agriculture. Let us briefly examine some aspects of agriculture that may help to put the potential socioeconomic effects of conservation tillage—as distinct from the ecological ones (though they are not unrelated)—into perspective.

In the United States, we have apprehended agriculture in three distinct ways: as a "way of life," as an industry, and as a profession. Farming as a "way of life" is, to a degree, the romantic view that *Newsweek*'s economics writer Robert J. Samuelson, in a 1985 column highly critical of U.S. farm policy, says is a "fixed image" in our public perception of agriculture, "even though farming is forever changing." Samuelson is correct. We cling to a nineteenth-century image of homesteading on the plains and of "mixed" farms with livestock, food crops, and cash crops on small acreages. Those who participated in this agriculture were those who could get to the land soonest—such as the Oklahoma "sooners." No other qualification was necessary except the ability not to starve, though many did, as the folk song "Starving to Death on a Government Claim" makes clear. The essence of this kind of agriculture was self-sufficiency; its goals were as much social as they were economic. But in time, after the first "tractor boom" of the 1920s and the plow-up

Part of the farm-policy problem, according to one journalist, is that we cling to a nineteenth-century image of agriculture. Those who participated in this agriculture were those who could get to the land soonest as it was opened up for agriculture, in the Nebraska sand hills in the photograph above, where many "starved to death on a government claim," or in the southern Piedmont.

of the fragile lands of the plains, it became clear to most observers that this kind of agriculture was producing anything but self-sufficiency. And it was ruining the land as well. When cash-crop prices went down, as they did all through the late 1920s but especially after 1929 (1932 farm income was one-third that of 1929), a revolutionary fervor took hold of the agricultural sector. The situation was frightening, and clearly something had to be done. When

Franklin Roosevelt began the Hundred Days at the outset of his presidency in March of 1933, the first order of business was to save the banks. The second was to save agriculture. As the head of the Farm Bureau warned Congress two months before FDR's inauguration, "Unless something is done, we will have revolution in the countryside within less than twelve months."

The history of agriculture during this period was incredibly rich, complex, and emotionally stirring. And the upshot was that while the rhetoric of the romantic, small-farm view was kept, policies were put in place that would shortly lead to the full industrialization of agriculture. It became conscious policy to reduce the number of farms and to make those that remained economically sound. It was a policy that rewarded mechanization, the use of new chemicals, and the consolidation of farmland into fewer ownerships. In return for industrialization, agriculture could count on some guarantees from the government—to protect prices, to eliminate foreign competition, and to develop an agricultural-research system with a clear-cut mission to increase productivity. The USDA, called "the people's department" by Abraham Lincoln, became, in effect, a trade ministry.

This has been our general approach to agriculture ever since. We have had a "two-agriculture" policy: lip service to small farms (way of life), dollars to large ones (industrial agriculture). But this did not take place without just cause, for the excess of small farms during the first third of this century ravaged one hundred million acres of cropland and triggered the greatest mass migration in American history.

Since that time, Vermont, as a small-farm state, has provided an extreme example of how these policies have worked over the long pull. Since World War II, Vermont lost two-thirds of its farms while the United States as a whole lost about half. But half is plenty. During one generation—the generation of 1950s' and 1960s' postwar affluence and rapid economic development—small farms all across the United States were going under at the rate of two thousand per week!

Some of these farms provided land for suburban housing and recreational development. Some were bought up by neighboring farmers. Some were acquired by corporations, real-estate speculators, hobby farmers, or those few who manage to be all three. Some farms were simply abandoned for unpaid taxes or foreclosed. At the rate of two thousand per week, whatever the reason, it became pretty clear that drastic changes were taking place in the working landscape. In 1950, some 15 percent of our population lived on a farm. Today it is only 2.2 percent. The question now is, will stability

During the era of post–World War II affluence, small farms were lost at the rate of two thousand per week, many of them to subdivisions such as this.

ensue, since most of the farms are gone anyway, or will new changes take place in what the farm-policy analysts in Washington, D.C. call "the structure of agriculture"?

□   □   □

To be sure, the rate of loss in numbers of farms has slowed, since there are fewer and fewer farms to be lost. But if those who make it their business to predict the future of the "structure" of agriculture are to be believed, that's small comfort to Americans who value and wish to protect a diversified farm economy and who wonder what effect, if any, the widespread adoption of conservation tillage will have in this regard. For them, surely one of the most disturbing government documents around is a 1980 report entitled, blandly enough, *U.S. Farm Numbers, Sizes, and Related Structural Dimensions: Projections to the Year 2000*, published in 1980 by William Lin, George Coffman, and J. B. Penn, all of the USDA's Economic Research Service. Future agricultural production, predict the authors, "will be dominated by fewer and fewer farms. By 2000, the largest one percent of farms will account for about half of all farm production. By contrast, fifty percent of the farms—the smaller ones—will produce only one percent." In the end, say Lin, Coffman, and Penn, half the farmland in the United States will be farmed by just fifty thousand farmers.

In an update of these findings, the U.S. Office of Technology Assessment issued a report in 1986 called *Technology, Public Policy, and the Changing Structure of American Agriculture*. In it, the OTA concludes that (1) small farms can no longer provide farmers an adequate family income; (2) "moderate-sized" farms are a stagnant sector in the structure of agriculture; and (3) the large and very large farms are increasingly dominating agriculture. Following is a table comparing current and projected statistics.

| | | 1982 | | 2000 | |
| | | Number of | | Number of | |
Class	Sales	Farms	Percent	Farms	Percent
Small and part-time	$20,000–$99,000	1,937,000	86	1,000,000	80
Moderate-sized	$100,000–$199,000	181,000	10	75,000	6
Large and very large	$200,000 and over	122,000	4	175,000	14

Source: U.S. Congress, Office of Technology Assessment. *Technology, Public Policy, and the Changing Structure of American Agriculture.* Washington, D.C.: U.S. Government Printing Office, 1986.

And so it would appear that commercial farming will become a livelihood for only a few very well-capitalized people. The average price of entry into that rarefied category of farms that will control the land, and therefore the farming business, in the year 2000 will be in excess of $2 million, according to Lin, Coffman, and Penn. They also suggest that the number of new farmers under thirty-five years of age will decrease by 40 percent by the year 2000, by which time "multiownership farms (corporations and partnerships) may account for half of all farm sales."

All of these findings go exceedingly against the grain in America, for they call into question a basic historical element in the national ethos: the right of an American to have access to land and, with hard work and sacrifice, to make a whole living growing something on it.

The "larger-and-fewer-farms" issue certainly offended Bob Bergland, perhaps our most interesting secretary of agriculture since Henry Wallace. Bergland, who served during the Carter administration, told a Farmers Union convention: "I for one do not want to see an America where a handful of giant corporations own, manage, and control the entire food system. Yet that is where we are headed if we don't act now."

Heeding his own warning, Bergland, a former congressman and Minnesota farmer, set up what is known as "the structure study"— as unflinching and controversial an examination as the Department of Agriculture has ever made of the industry that it serves. The title of its report, issued in 1981, is *A Time to Choose: Summary Report on the Structure of Agriculture.* It decried mechanization, bigness, technological overkill, and the failure of resource steward- ship. But evidently, we had already chosen: few people heard of the report; it had no discernible impact. In fact, it is not even possible to get a copy from today's Department of Agriculture. (Just to check, when the department's general number was called, the caller was passed through to the "Subject Matter Section." The woman answering there said: "Oh, *that* report. That was done by a previous administration. We wouldn't have any copies of *that.*"

Like all government reports, *A Time to Choose* is no literary land- mark. A supreme act of will is required just to sit down and read it—except for its opening section, a quite personal introduction by Bob Bergland himself. It begins: "I grew up on a farm that my grandfather bought eight years before I was born, just before the World War I 'boom' broke. He, my father and my uncles worked that farm through those bad years. At that time 27 percent of all Americans gainfully employed were farming."

Pointing out that American agriculture has provided extraordi- nary abundance over the years, Bergland was concerned, for the most human of reasons, that the structure was working against the "idea" of farming in the United States, even though farming as a whole was so tangibly successful: "Whenever I wondered aloud if we were on the right policy and program track, I was pointedly reminded that abundance is the main objective of the system, that this had been the goal of farm legislation for 50 years. But what had happened to our farm system along the way?"

What had happened was that it had changed into an "industry." The very difficulties that Bergland and his crew of young idealists in the Carter administration's Department of Agriculture found so worrisome had, in fact, been exactly intended by policies created by the "brain trust" in FDR's first term. Brain truster Rexford Tug- well, an architect of the New Deal farm program, wrote in 1936 that the most important part of the new farm policy was to "increase the productivity of the American family farm. This has been par- tially the effort to make 'two blades of grass grow where one grew before,' but only partly that. Most signficant has been the making of more blades of grass to grow for one farmer. We have not increased production per acre nearly as fast as we have increased production per worker."

Tugwell needn't have worried. During the crucial post-war period, in which great changes in the business of agriculture have taken place, farm output has very nearly doubled while "input," including acreage under cultivation, has remained virtually the same, with overall productivity increasing by two-thirds. These enormous gains in productivity were the result of economies of scale interacting with irrigation, chemical fertilizers, herbicides and pesticides, new, high-yielding genetic strains, new approaches to livestock management featuring feedlots rather than allowing animals to range about on the wide-open spaces, and the vertical integration of agriculture in many sectors (managing the entire food production and marketing process from planting to packaging and promotion on the supermarket shelves in order to maximize profit all along the line), together with vast improvements in transportation to move produce swiftly from farm to market.

As recounted in Chapter 2 of this book, the fantastic increase in productivity and income potential coincided, neatly enough, with

Demonstrations staged by the American Agriculture Movement in 1979 to provide greater support to an industry already in trouble did have an effect, though somewhat belated. Between 1982 and 1986 the Reagan administration has pumped more than $90 billion into the farm economy. But even that astonishing subsidy hasn't worked.

an increase in foreign markets for our commodities—beginning with that most dramatic event, the 1973 Russian wheat deal, wherein a substantial fraction of U.S. grain was sold to the USSR for less than market value, which drew down reserves to the point that bread became nearly as dear as cake. By the end of the decade, agricultural exports had exceeded $40 billion per year, and the United States became "the OPEC of soil." Counseled Secretary of Agriculture Earl Butz at the beginning of the foreign-market boom, "Plant fencerow to fencerow." In a cover article on the Butzian philosophy, *Time* magazine warned farmers, via its headline, to "get big or get out." Many took the advice. They got out. Perhaps they were the lucky ones, because those who borrowed money at high interest rates in order to "get big" wound up flat broke. It was these farmers who, organized as The American Agriculture Movement, came to Washington (after Butz was gone and Bergland was trying to pick up the pieces) with their dual-wheeled diesel tractors—bought on time—and demanded redress.

If AAM demonstrators were disappointed then, they had only to wait a few years. Between 1982 and 1986, the Reagan adminstration pumped over $90 billion into the farm economy. But even that astonishing subsidy hasn't worked. After the billions upon billions spent by the Department of Agriculture, the result is still foreclosure for many, even if a few are saved by the government programs entrained by Rexford Tugwell and his colleagues so long ago. For those who remain in the farm sector—that 2.2 percent of the population—a sense of beleaguerment prevails, with oft-repeated apothegms spoken at the kitchen tables at day's end: "You know, mother, us and the kids could've done better if we'd just sold out and put the money in the bank to draw interest."

                              □   □   □

Just as the way-of-life concept was taken to extremes in the first third of the century, resulting in so much tragedy, so the agriculture-as-industry concept replacing it seems to have exceeded its limits, its failure hastened and made vivid in the great export-boom period, 1973–1982. The boom caused an extraordinary overextension in agricultural debt for the purchase of machinery and land based on the premise that a 40 percent level of exports would continue forever. It did not continue forever, and while many hope that someday the United States may yet recapture some overseas markets for its farmers, many others fear that it may not.

In the fall of 1985, a *New York Times* reporter, James Sterngold, visited the famed commodity pits of the Chicago Board of Trade, where high rollers standing hip to haunch once shouted bids at one another for futures in grain, soybeans, pork bellies, and live cattle in an exciting, swirling cacophony that dramatized the vitality, exuberance, and abiding strength of our greatest American enterprise, agriculture. Ordinarily, fall would be high season for commodity trading, but now the scene was quite different. "The small amphitheatre-shaped pit where corn options are traded," Sterngold said, "was almost deserted." In 1980, nearly twelve million contracts had been traded in this pit. The volume in 1985 was scarcely half that amount. As one floor trader told Sterngold, "There have been days when I have not made one trade."

The eerie quiet of the pits is richly symbolic of the present state of American agriculture. There have been slumps before, of course, but this one may well be different. The United States has lost its dominance as a major commodity exporter. Our share of the world market for wheat, for example, went from 50 percent in 1980 to 26 percent five years later. In May 1986, American agricultural export sales were actually less than agricultural import purchases—for the first time since 1971. The deficit for the month was nearly $350 million, an amount, according to an account in the *New York Times*, that set some economists to speculating that the United States might become a net food *importer* on an annual basis by 1990.

The conventional reasons given for the agricultural recession are that (1) the 1979 grain embargo undermined our reliability as an exporter; (2) the high price of the dollar versus other currencies

The loss of export markets, which may be a permanent loss, has caused enormous surpluses of commodity crops. The final stage of the industrialization of agriculture has been reached. Some believe that it's time for a major change in policy.

has made U.S. farm exports too expensive; (3) high bank interest rates have created a volatile cost-price situation.

All of these reasons seem plausible. Therefore, it should follow that after policies on embargoes are clearly changed (which Reagan has done, reversing Carter), after the dollar loses some value on currency exchanges (as it has done; too quickly for some), and after bank interest rates come down (they have, dramatically), *then* everything will stabilize: markets will expand abroad, and farming will become profitable again.

But the road back to this kind of agricultural prosperity has another twist in it that is not often included in these kinds of analyses. Multinational agribusinesses, operating in Third World countries and using the breakthroughs provided by the Green Revolution along with low-cost land, labor, and capital, are moving into world markets formerly dominated by U.S. farmers. Soybeans from Brazil can be sold in Chicago cheaper than those from downstate Illinois. Only "Section 22," a protectionist measure enacted in 1935 as an amendment to the Agricultural Adjustment Act, keeps the multinationals from competing for domestic sales. Like Catch-22, Section 22 effectively conceals what may really be happening: the permanent loss of a world market that just a few years ago was buying the produce of two U.S. cropland acres out of five.

And so the price of farm acreage declines—except in those areas where values are influenced by urban expansion—and borrowing power declines with it. Land prices are now half what they were in the heart of the great commodity districts, leaving landowners with half the borrowing power they once had—at a time when they need more credit, not less, just to stay even on the commodity treadmill in a declining market. As of 1986, the lender of last resort, the Farm Credit System, is facing losses of up to $6 billion on its $73 billion portfolio, despite giant farm subsidies a whole order of magnitude higher than they were during the decade of the 1970s.

Thus does the curtain threaten to fall on the age of heavy-industry-style agriculture, just as it did on way-of-life farming fifty years ago. The end point of agricultural industrialization, which was to bring about economic and social stability, has produced just the opposite. "For half a century," *Newsweek*'s Samuelson writes, "we have subsidized and pampered farming. By any test, these policies have failed. . . . They have not stabilized farm incomes. And they have not expanded food exports; indeed in recent years, they've done just the opposite."

□   □   □

It takes no great foresight to realize that farm policy in the United States is once again on the cusp, perhaps about to go through as massive a change as it did earlier in this century. Certainly, the way we are thinking about farms and farming is changing, as journalists like Samuelson amply demonstrate. So we come to the third way we can apprehend farming: agriculture as a profession. And this, as it happens, has a great deal to do with conservation tillage, which may be seen as both a cause and an effect of a new, and quite urgent, emphasis on professionalism in agriculture. It is a cause for professionalism since it tends to favor the expert farmer and reward him with increased profits—not as a result of his capital expenditures or his blind luck in inheriting land (though it helps), but as a result of his skills. Conservation tillage is an effect of professionalism because it was created by professional farmers in the first place, unwilling to tolerate the damage to the soil caused by the moldboard plow and yet aware of the impossibility of incorporating expensive conservation countermeasures at a time of rising costs and lowering prices.

Thus, conservation tillage comes along at a fortuitous moment. No longer—for all the reasons set forth in the preceding pages—can agriculture be a matter of who got on the land the soonest, as in the nineteenth century. No longer can it be a matter of who owns the biggest tractors and the most land, as it has been for the past fifty years. The economy of a nation that depends on its agricultural land-resource base as much as does ours now must insist that agriculture be concerned with who can *farm the best*. And while there is nothing new about the concept of professionalism in agriculture, it has, remarkably, never been encouraged—either by policy, which has gone for bigness, or by public sentiment, which still romanticizes farming along nineteenth-century lines.

Perhaps the most eloquent modern spokesman for professionalism in agriculture was Louis Bromfield, the novelist-farmer who in 1939 founded Malabar Farm in Ohio, the most celebrated demonstration farm in agricultural history. Often dismissed as a crank by the experts of his day (they were deeply committed to industrializing agriculture, remember), Bromfield, like his fellow Ohioan Edward Faulkner, believed in scientific farming—going about agriculture in a fine-grained, inquiring way rather than through the thoughtless application of massive amounts of horsepower to larger and larger acreages. If the experts were skeptical, farmers loved it. One field day at Malabar attracted eight thousand visitors.

But what attracted farmers to Bromfield was not just his experiments, but also his insistence on professionalism and on the worth

Louis Bromfield, novelist, political iconoclast, and owner of Malabar Farm, perhaps the most famous experimental farm in the country, claimed that subsidies were no answer to an ailing farm economy. He was a proponent of professionalism in farming, long thought to be desirable but so far unsupported by government policy. Conservation tillage may encourage the kind of professionalism Bromfield celebrated.

of farming as a profession.[1] As he wrote in his book *Out of the Earth* (1950):

> Perhaps no stupider human saying has ever been formulated than the one that "anybody can farm." Anyone can go through the motions, but not 10 percent of our agricultural population today could seriously be called "good farmers." Thirty percent are pretty good, and the remaining 60 percent do not, through ignorance or laziness or sometimes through the misfortune of living on wretched land fit only for forests, deserve the dignified title of "farmer." Most of them still remain within the range of a completely primitive agriculture confined to plowing, scattering seed and harvesting whatever crops with luck turn up at the end of the season. That they perform these operations with the aid of modern machinery does not make them either good or modern farmers. Tragically, a great many of them actually hate the soil which they work, the very soil which, if tended properly, could make them prosperous and proud and dignified and happy men.

---

[1]Bromfield also decried farm subsidies, believing them to retard rather than advance economic stability in agriculture. Only now are mainstream farm analysts beginning to agree with him.

Bromfield was a latter-day Jeffersonian, for Jefferson too believed deeply in the importance of a professional agriculture—and was a professional himself. While Jefferson championed the idea of "a little portion of land" for any American willing to work it, this did not mean that he favored the emergence of an American peasantry with a land-reform scheme that would put them to work on tiny plots. His model was the British yeomanry, a rural middle class that arose in Elizabethan England and that had achieved some political power in Britain by Jefferson's time. The successful yeoman farmer became a "statesman": a landowner with a good, big house and great respectability and dignity; a community leader; a man of culture and learning. In short, a farmer.

Now, a new kind of agriculture has come along that puts a premium on the very qualities that Bromfield, and Jefferson, and all the people interviewed for this book, and the author himself believe are essential for the "structure" of American agriculture to achieve. This is the larger promise of conservation tillage. In our national discussion of agriculture, we have tended to view policies as either-or: either for the small, family farm or for the large, industrial farm. Either for careful, *intensive* farming techniques or for big-acreage, *extensive* techniques. Either for "mixed" farms with many crops and long rotations or for high-tech, single-crop monocultures. No wonder our policies have become so distorted. Now, the time has come for a paradigm shift in our way of thinking, and conservation tillage may lead us to it because it obviates an old duality that no longer has any basis in fact.

Conservation tillage is, after all, a *front-edge* technology, attractive to young people, particularly bright young people. Do not think of a conservation-tillage farmer as "Farmer Jones," white beard speckled with hayseed. Think of Todd Greenstone, child of the upper middle class in America who refused to become a faceless yuppie in a big-city office tower. Bright as he is handsome, he walks through his fields in chinos and a sport shirt, and back in his office, calculator in hand, computer at the ready, he "farms the desk." A young *rural* professional. And he didn't even have to own any land to do it. (Though he will; count on it.)

Conservation tillage may even rejuvenate the "moderate-sized farm" category that the Office of Technology Assessment has claimed is moribund. While conservation tillage assesses no particular penalities for "bigness," it may tend to encourage a renascence of the moderate-sized commercial farm, operated by a single family, as the most efficient and most economically prudent to own and manage. The fact is, the economies of scale peak at a lower level of investment for conservation tillage compared with the extremely capital-intensive farming it replaces. The Eppleys' farm is of moderate size, and the couple now farm it themselves, no outside labor necessary. The economy of conservation tillage is concerned more with "net" than with "gross." The economic advantage of a large number of acres is much reduced. And the effect of these factors is to produce a quite different life-style, as Rosemary Eppley so eloquently points out. The Eppleys are much in demand as lecturers throughout the Midwest.

Conservation tillage makes for a creatively challenging agriculture. The work of Morton Swanson and Jerrell Harden seems remarkable today—almost Jeffersonian. The intellectual and scientific achievement of both these men is extraordinary. But in times to come, perhaps their inquiring and inventive approach to farming will come to be expected of the farm professional, just as it is for professionals in other fields—lawyers, physicians, CPAs. Why shouldn't farmers, rather than faculty members at land-grant colleges, be writing the articles for learned journals? Perhaps in time, the technical literature of conservation tillage will be primarily, instead of only partially, farmer-written. It would no doubt be the better for it.

And, finally, conservation tillage may encourage farms to stay put. The exodus is over. The earthworms have come back to the Eppleys' cornfields. Erosion has been halted on Mort Swanson's steep wheat lands. Generations out of mind can make a good living from the Harden's sandy, red soil. Tilth is abuilding on the rented fields of Todd Greenstone.

Will the dire projections about the "structure of agriculture" winding up like the structure of, say, the steel business have to be modified? No promises can be made about that yet. But still, professional conservation tillage might well increase the number of full-time farmers on farms of "moderate size." The cost of entry for young farmers may be less than projected. More farmers, rather than fewer, can be in the "most productive" category. And because indebtedness can be reduced and costs controlled, conservation tillage can provide a more reliable living, and with more discretionary time into the bargain.

When the massive federal subsidies to agriculture are withdrawn, and surely they will be sooner rather than later, the moderate-sized conservation-tillage farmers—given lower acreage requirements, lower "input" costs, and high levels of discretionary time—may well be able to weather the change better than most. And in a farm economy finally more responsive to actual market forces, they are the ones most likely to prosper. Moreover, with a permanent investment made in the quality of the soil by conser-

Conservation tillage comes at a time when stability is desperately needed in farming, especially for owner-operators of medium-sized farms like this one in Carroll County, Maryland. Thomas Jefferson believed that farming should be economically attractive and that it is a dignified way of life, expressive of the best in the American character. Perhaps the American "tillage revolution," like the earlier one Jefferson had so much to do with, can bring us closer to that goal.

vation-tillage farmers, geographical stability in agriculture will be not only possible, but necessary.

□   □   □

Now for a final word. This book has no high didactic purpose. It is not written to encourage conservation tillage, although it is written in praise of it. It is not written to provide an agenda for soil conservation, or environmental quality, or social justice. And yet all those things are implied by the "tillage revolution." This is a moment of pause in the growth of conservation tillage. It is well begun, and "phase two" is on the way, a phase that may be characterized in part by technological innovation but also in part by a paradigm shift in the way we apprehend important issues regarding the structure of agriculture, rural life, and the whole economy of the United States. In these matters, conservation tillage can provide a fulcrum, an organizing principle, to help meet these challenges creatively. It is essential for this reason that all citizens, not just farmers and those who work with them, know what conservation tillage is and what it implies.

Those who have read all the way to this point have a bit of understanding. The next step is to do what the author did: call up Todd Greenstone, or Mort Swanson, or Carl and Rosemary Eppley, or Jerrell Harden and his dad and son—or whomever you can find who is a conservation-tillage farmer—and go visit them. You will learn about conservation tillage. And you will learn something about the heart of America: its land and farms, its people, and its hope.

Postscript: A Personal Note

This book has turned out to be much more of a celebration of conservation tillage—as opposed to a critique—than I, as a writer on natural resources and environment, originally intended. Believe me, it didn't start out that way. Let me explain.

Several years ago—1981 or so—when Neil Sampson, a friend of mine, was head of the National Association of Conservation Districts, NACD organized the Conservation Tillage Information Center, which was to provide technical information to farmers on how to go about giving conservation tillage a try. I was renting some office space from NACD at the time so was keeping up with their effort to get the CTIC funded and off the ground. I had been fascinated by conservation tillage since the mid-1970s when I first wrote about it for the Congressional Research Service of the Library of Congress. But when I learned that the largest part of the early funding for the information center was from chemical companies, I was . . . well, *appalled* is not too strong a word. Was the information conveyed by the center to be how terrific herbicides were? Scarcely a less disinterested source of support could be found for a tillage system that the Orthos, Monsantos, and DuPonts already thought they owned lock, stock, and grain drill.

Sampson was, and is (he has since moved on to become executive vice president of the American Forestry Association), one of those rare conservationists who can keep a foot in the resource *users'* camp as well as that of the resource preservationists. I try to do this myself, but tend to drift over to the preservationist side, which is the side that does not love herbicides. Accordingly, I asked Neil, more or less at the top my voice, whether NACD's acceptance of chemical-industry money to support an otherwise good idea wasn't a case of the fly not only acceding to the spider's invitation into her parlor, but hopping right up on the dinner table just to be obliging? Neil answered that if I would just sit down and be quiet, he would explain a few things to me.

The first thing he explained was that there would *be* no conservation tillage without herbicides. That is, of course, factually correct, but its wisdom is tediously conventional. Then Neil told me something else, which at the time I thought was an overly subtle point. He said that conservation tillage in its present form should be seen as a *pathway*, not as an end-state technology. He had a sense, which he found hard to articulate, that there was something larger going on in conservation tillage than the use of herbicides. Larger even than the immediate gains conservation tillage could promise in limiting soil erosion and non-point-source pollution. He said that it was important to promote conservation tillage to farmers so that it could get to wherever it was going, even if its proponents had to strike a bargain with chemical companies to do so. "We are inventing, or reinventing," Neil told me, "a new kind of interrelationship between crops and soils." I responded that I was glad to hear it, but that I didn't know what he was talking about.

A few years later I wrote an article for *Country Journal*, some paragraphs of which have survived in Chapter 2 and Chapter 7. While supportive of conservation tillage, the article questioned its reliance on herbicides. Then the opportunity to write this book came along, and at first I thought it would be a simple expansion of my article—getting at the herbicide issue in some detail. But halfway through the field research, I discovered what Neil meant. Something *historical* was happening in the way farmers and agronomists were thinking about tillage. Herbicides were a large part of the picture but certainly not the definitive motif that I thought. In this book, I have tried to put conservation tillage into a context that is consonant with this finding. I take nothing back from my feelings concerning the present dependence of conservation tillage on herbicides. And though I am much relieved at progress in herbicide development and use as detailed in Chapter 7, it seems to me that current industry efforts to bioengineer crop cultivars to be resistant to them is grossly retrograde, an Edsel of an idea and deserving of a similar fate. Nevertheless, to see conservation tillage, as many of my colleagues in the environmental movement do, simply in terms of herbicides versus erosion is to miss the basic ecological point of it all—the idea of maintaining the natural integrity of the soil.

Why was it that I was so slow to get the ecological message of conservation tillage? One reason is a dearth of *usable* published information that gives this important topic any meaningful context at all. The CTIC itself is not, despite its name, a source of public information, but of quite specific technical data of interest mainly to practitioners, government officials, and agronomists. Published sources are pretty much limited, on the one hand, to the breathless, gee-whiz journalism of farm magazines or booklets published by them or by the public-relations departments of agribusinesses, with their chatter about bushels and dollars, and on the other hand, to academic articles of such monumental opacity that they seem almost like obscure theological exegeses of medieval monks. The most interesting materials, I found, are the often *self*-published manuals by farmers themselves—such as Ernest Behn's book, from which I quoted liberally in Chapter 5. Farmers understand the meaning of conservation tillage rather better than academics or agricultural journalists. But even their information is "inside baseball," not easily decipherable by a nonfarmer. As a result, I wound up relying most heavily on personal interviews, which I tape-recorded. When I asked farmers and the agricultural experts who work most closely with them to be interpretive about conservation tillage, they were able to do so. The cassettes fill up the better part of a shoe box. But this is not a body of information that exists in published form, except on very much of a here-and-there basis.

It was, I suppose, to be expected. Farming is the foundation of the largest segment of our economy, but all the same, our most diffuse industry. There are no central, well-organized communications instrumentalities by which professional farmers can explain themselves to the general public as, for example, do home builders or physicians. I should be quick to point out that there are plenty of communications instrumentalities controlled by agribusinesses—those who purvey the "inputs" to farmers. Moreover, there are organizations of farmers with elaborate informational capacities, but for the most part, these are to provide services to their own membership or constituencies (the CTIC is an example) or to agitate for more favorable government policies. I know that the federal government tries to provide public information on farming, and I am grateful for the many services they have provided me, especially the USDA's Agricultural Research Service. But for the average citizen, there are so many voices and so many conflicting points of view from government sources that for the most part, the result is noise rather than communication.

It is tempting to recommend the establishment of, say, a national agricultural-science information bureau—a kind of organized shoe box—as the answer to the lack of interpretive information on various technical aspects of farming. But my hunch is that the world of agriculture does not need any more "PR." Instead, the root of the problem, at least as I perceive it in myself, is the abysmal lack of agricultural education in this country. The reason the agriculture industry has no interpretive information on conservation tillage to speak of is that the public does not know how to ask for it. We do not know the terms of agriculture, the language, the basic concepts. And outside of taking vocational courses or actually working on a farm (which lets 98 percent of us out), there simply is no way for anyone readily to find out what the basic concepts are.

Thomas Jefferson, who figures large in this book, believed agriculture should be part of a liberal college education: a science to be studied along with, say, chemistry. College students study other sciences without a vocational intention. More than mechanical engineers study physics; you do not have to sign on as a zookeeper to study zoology, or become a miner to take geology. But if you do not intend to become a farmer or an agronomist, there is no real way for you to study agriculture.

The point is that agriculture, like physics, zoology, and geology, is, I have come to believe, worthy of study for its own sake, as a science. It seems to me that at bottom, our (okay, then, *my*) agricultural ignorance is not a matter of a failure of PR—we cannot hold the farmers themselves responsible for it. Instead, it must be addressed fundamentally through formal scientific education, quite aside from the grammar-school business of field trips to visit Farmer Jones or high-school social-science "units" on farming. When you go to college and have to select courses from Group III (as the sciences were designated in my school), agricultural science ought to be among them. And for the first lecture in Ag Sci 101, the professor should start with the soil: the physics and chemistry and biology of the soil. That's where agriculture begins. All the rest, to borrow a phrase from the Talmud, is commentary.

And so it is that I have emphasized the ecology of the soil throughout this book as the lens through which conservation tillage should be seen and understood. For those who worry about where an author is "coming from," that's where I am. I wound up in full agreement with my friend Neil Sampson after all; maybe I even took his insight a step further. But I'll confess (bad word, but I can think of none other) that getting there wasn't easy.

List of Principal Sources[1]

Interviews

Christiansen, Ralph. U.S. Environmental Protection Agency. Chicago, Illinois. (t)

Cook, Ken. Conservation Foundation. Washington, D.C.

Craig, Marvin. Environmental writer. Glendale, California. (t)

Dickerson, Chester. Monsanto Corporation. Washington, D.C. (t)

Dillman, Don. Department of Rural Sociology, Washington State University. Pullman, Washington.

Doyle, Jack. Environmental Policy Institute. Washington, D.C. (t)

Dunsmore, George. Vermont Department of Agriculture. Montpelier, Vermont.

Ebenreck, Sara. National Association of Conservation Districts. Washington, D.C.

Elkins, Charles B. National Soil Dynamics Laboratory. Auburn University, Auburn, Alabama.

Eppley, Carl and Rosemary. Farmers. Wabash, Indiana.

---

[1]Interview entries show affiliations at the time of interview. If an entry is followed by a (t), this means that the interview was conducted by telephone. All others were conducted in person. Published sources that may be of special interest for readers wishing more information on conservation tillage are starred (*), and where appropriate, publishers' addresses given, along with a comment or two.

Fletcher, Donna. U.S. Environmental Protection Agency. Washington, D.C. (t)

Greenstone, Todd. Farmer. Brookville, Maryland.

Griffith, Donald R. Cooperative Extension Service. Purdue University, West Lafayette, Indiana.

Harden, J. C. Farmer. Banks, Alabama.

Harden, Jerrell. Farmer. Banks, Alabama.

Harden, Leo. Bush Hog Manufacturing Company. Selma, Alabama. (t)

Hinkle, Maureen. National Audubon Society. Washington, D.C. (t)

Kaufman, Donald D. Agricultural Research Service, U.S. Department of Agriculture. Beltsville, Maryland. (t)

Kerr, Stephen. Vermont Department of Agriculture. Montpelier, Vermont.

Lake, James. Conservation Tillage Information Center. Fort Wayne, Indiana.

Luth, Goeffrey. Yielder Drill Company. Spokane, Washington.

Madden, Patrick. Department of Agricultural Economics. Pennsylvania State University, University Park, Pennsylvania. (t)

Moldenhauer, William C. National Soil Erosion Laboratory. Purdue University, West Lafayette, Indiana.

Ogg, Alex. Agricultural Research Service. Washington State University, Pullman, Washington.

Papendick, Robert I. Agricultural Research Service. Washington State University, Pullman, Washington.

Rasmussen, Wayne. Agricultural Historian, U.S. Department of Agriculture. Washington, D.C. (t)

Reeve, Wayne. National Soil Dynamics Laboratory. Auburn University, Auburn, Alabama.

Riggleman, James. E. I. DuPont Company. Wilmington, Delaware. (t)

Rogers, Ken. Soil Conservation Service, U.S. Department of Agriculture. Auburn, Alabama.

Schertz, David. Soil Conservation Service, U.S. Department of Agriculture. Washington, D.C. (t)

Schmick, Greg. Yielder Drill Company. Spokane, Washington.

Schnepf, Max. Soil Conservation Society of America. Ankeny, Iowa. (t)

Schwartz, E. Bruce. Agricultural Research Service, U.S. Department of Agriculture. Beltsville, Maryland. (t)

Shafer, Karen. U.S. Environmental Protection Agency. Washington, D.C. (t)

Steiner, Frederick. Department of Horticulture and Landscape Architecture. Washington State University, Pullman, Washington.

Swanson, Morton. Farmer. Colfax, Washington.

Taylor, James H. National Soil Dynamics Laboratory. Auburn University, Auburn, Alabama.

Wagonner, Peggy. Rodale Research Center. Kutztown, Pennsylvania. (t)

Weller, David. Agricultural Research Service. Washington State University, Pullman, Washington.

Williamson, Gus. Yielder Drill Company. Spokane, Washington.

Youngberg, I. Garth. Institute for Alternative Agriculture. Beltsville, Maryland. (t)

Books

*Behn, Ernest E. *More Profit With Less Tillage*. Boone, Iowa: Behn Enterprises, 1982. An engaging account of one farmer's experience with conservation tillage. A must for anyone interested in "ridging." Available directly from the publisher at Route 1, Boone, Iowa 50036.

Berry, Wendell. *The Unsettling of America: Culture and Agriculture*. New York: Avon Books, 1977.

*Betts, Edwin Morris. *Thomas Jefferson's Farm Book*. Princeton, New Jersey: Princeton University Press, 1953. Facsimile reprint of T. J.'s farm notebook (there's a companion *Garden Book* dealing with horticulture) plus, best of all, a fat selection of Jefferson's letters and other papers concerning agriculture. Whoever owns this wonderful piece of scholarship can count themselves lucky, for the book is out of print. If you want to sell a copy, contact the author.

Bowers, William L. *The Country Life Movement*. Port Washington, New York: Kennikat Press, 1974.

Bromfield, Louis. *Malabar Farm*. New York: Harper & Brothers, 1948.

*Bromfield, Louis. *Out of the Earth*. New York: Harper & Brothers, 1950. If only one Bromfield book is to be read, for those interested in the experiments at Malabar and the author's farm philosophy, this summary work is probably the best bet. Out of print, but a reprint edition is available from Amereon Ltd., Box 1200, Mattituck, New York 11952.

Bromfield, Louis. *Pleasant Valley*. New York: Harper & Brothers, 1945.

Carson, Rachel. *Silent Spring*. Boston: Houghton Mifflin Company, 1962.

Dubos, Rene. *A God Within*. New York: Charles Scribner's Sons, 1972.

*Faulkner, Edward H. *Plowman's Folly*. Norman, Oklahoma: University of Oklahoma Press, 1943. Out of print, but many libraries have it.

Faulkner, Edward H. *A Second Look*. Norman, Oklahoma: University of Oklahoma Press, 1947.

Fletcher, W. Wendell, and Charles E. Little. *The American Cropland Crisis*. Bethesda, Maryland: American Land Forum, 1982.

Goodwyn, Lawrence. *The Populist Movement*. New York: Oxford University Press, 1978.

*Hayes, William A. *Minimum Tillage Farming*. Brookfield, Wisconsin: No-Till Farmer, Inc., 1982. An excellent handbook for ridge tilling and associated techniques, with lots of tables and worksheets useful to the farmer. Publisher's address is P.O. Box 624, Brookfield, Wisconsin 53005. Note: this publication is produced in a maddening double-book format with another book upside down and back-end to it. See citation under Young.

Ivins, Lester S., and A. E. Winship. *Fifty Famous Farmers*. New York: Macmillan, Inc., 1925. See especially the chapter on John Deere, pp. 21–30.

Jackson, Wes. *New Roots for Agriculture*. San Francisco: Friends of the Earth, 1980.

Klein, Joe. *Woody Guthrie: A Life*. New York: Alfred A. Knopf, Inc., 1980.

Koch, Adrienne, and William Peden. *The Life and Selected Writings of Thomas Jefferson*. New York: Modern Library, 1944.

Leakey, Richard E. *People of the Lake: Mankind and Its Beginnings*. New York: Doubleday & Company, Inc., 1978.

MacLean, Kenneth. *Agrarian Age*. New Haven, Connecticut: Yale University Press, 1950.

Malone, Dumas. *Jefferson and the Ordeal of Liberty*. Boston: Little, Brown & Company, 1962.

Malone, Dumas. *Jefferson and the Rights of Man*. Boston: Little, Brown & Company, 1951.

Malone, Dumas. *Jefferson the Virginian*. Boston: Little, Brown & Company, 1948.

Marx, Leo. *The Machine in the Garden: Technology and the Pastoral Ideal in America*. New York: Oxford University Press, 1964.

Morison, Samuel Eliot. *The Oxford History of the United States*. New York: Oxford University Press, 1965.

Morrissey, Charles T. *Vermont: A Bicentennial History*. New York: W. W. Norton & Company, Inc., 1981.

Rice, Robert W. *Fundamentals of No-Till Farming*. Athens, Georgia: American Association for Vocational Instructional Materials, 1983.

*Sampson, R. Neil. *Farmland or Wasteland: A Time to Choose*. Emmaus, Pennsylvania: Rodale Press, 1986. First-rate, comprehensive analysis of agricultural land-resource issues.

Schlesinger, Arthur M. *The Coming of the New Deal*. Boston: Houghton Mifflin Company, 1959.

Stratton, Joanna L. *Pioneer Women: Voices from the Kansas Frontier*. New York: Simon & Schuster, Inc., 1981.

Strong, Josiah. *Our Country*. (1886) Cambridge, Massachusetts: Belknap Press, 1963.

Thoreau, Henry David. *Excursions*. (1863) New York: Corinth Books, 1962.

Van Dersal, William R. *The Land Renewed: The Story of Soil Conservation*. New York: Henry Z. Walck, Inc., 1968.

*Worster, Donald. *Dust Bowl: The Southern Plains in the 1930s*. New York: Oxford University Press, 1979. Absolutely tops. Absolutely essential as background to conservation policy.

*Young, H. M. *No-Tillage Farming*. Brookfield, Wisconsin: No-Till Farmer, Inc., 1982. This is *the* farmer's handbook on no-till, first published in 1973. The author is the first commercial no-till farmer. Publisher's address: P.O. Box 624, Brookfield, Wisconsin 53005. Note: this book is published in an awkward format with another book; see citation under Hayes.

Research Monographs and Government Reports

Battelle Institute. *Agriculture 2000*. Columbus, Ohio: Battelle Press, 1983.

*Bauder, J. W. *Annotated Bibliography of Selected Extension Publications on Conservation Tillage*. Fort Wayne, Indiana: Conservation Tillage Information Center, 1984. For local research on conservation tillage, this guide is indispensable. Available directly from the Center, 2010 Inwood Drive, Fort Wayne, Indiana 46815.

Conservation Tillage: Local Action and Views, A Survey of America's Conservation Districts. Fort Wayne, Indiana: Conservation Tillage Information Center, 1984.

Cook, R. J., and W. A. Haglund. *Pythium Root Rot: A Barrier to Yield of Pacific Northwest Wheat.* Research Bulletin XB0913. Pullman, Washington: Agricultural Research Center, Washington State University, 1982.

Cornucopia Project. *Empty Breadbasket? The Coming Challenge to America's Food Supply and What We Can Do About It.* Emmaus, Pennsylvania: Rodale Press, 1981.

Cousins, Peter H. *Hog Plow and Sith: Cultural Aspects of Early Agricultural Technology.* Dearborn, Michigan: Greenfield Village and Henry Ford Museum, 1973.

Galloway, H. M., D. R. Griffith, and J. V. Mannering. *Adaptability of Various Tillage-Planting Systems to Indiana Soils.* West Lafayette, Indiana: Cooperative Extension Service, Purdue University, 1984.

Griffith, D. R., et al. *A Guide to Till-Planting for Corn and Soybeans in Indiana.* West Lafayette, Indiana: Cooperative Extension Service, Purdue University, 1982.

Holden, Patrick W. *Pesticides and Groundwater Quality: Issues and Problems in Four States.* Washington, D.C.: National Academy Press, 1986.

*King, Arnold D. *Conservation Tillage: Things to Consider.* Washington, D.C.: U.S. Department of Agriculture, 1985. An excellent technical introduction to conservation tillage. Ask for Agriculture Information Bulletin 461 from the Office of Public Affairs, USDA, Washington, D.C. 20250.

Lake Erie Conservation Tillage Demonstration Projects: Evaluating Management of Pesticides, Fertilizer, Residue to Improve Water Quality. Chicago, Illinois, and Washington, D.C.: Great Lakes National Program Office, U.S. Environmental Protection Agency and the National Association of Conservation Districts, 1985.

Lin, William, George Coffman, and J. B. Penn. *U.S. Farm Numbers, Sizes, and Related Structural Dimensions: Projections to the Year 2000.* Washington, D.C.: U.S. Department of Agriculture, Economics, Statistics, and Cooperatives Service, 1980.

Madden, Patrick. *Debt-Free Farming is Possible.* University Park, Pennsylvania: Cooperative Extension Service, Pennsylvania State University, 1985.

Mannering, J. V. *Conservation Tillage to Maintain Soil Productivity and Improve Water Quality.* West Lafayette, Indiana: Cooperative Extension Service, Purdue University, 1979.

National Survey of Conservation Tillage Practices (1985). Executive Summary. Fort Wayne, Indiana: Conservation Tillage Information Center, 1986. See Appendix I for excerpts.

Obert, John C., and William A. Galston. *Down, Down, Down on the Farm.* Washington, D.C.: Roosevelt Center for American Policy Studies, 1985.

Paddock, Joe, Nancy Paddock, and Carol Bly. *The Land Stewardship Project Materials.* St. Paul, Minnesota: The Land Stewardship Project, n.d. (1982?)

*Sand, Duane. *Resourceful Farming Through Conservation Tillage.* Des

Moines, Iowa: Iowa Natural Heritage Foundation, 1986. A nifty little report, nicely turned out as a booklet, summarizing what conservation tillage is and what can be expected of it from an environmental conservationist's point of view. Available from the Foundation, Insurance Exchange Building, Suite 830, 505 Fifth Avenue, Des Moines, Iowa 50309.

*Seneca County No Till and Ridge Till Demonstration Results, 1984. Tiffin, Ohio: Seneca Soil and Water Conservation District, 1985. An excellent model study. Available from the district office at 155 East Perry Street, Tifflin, Ohio 44883.

Six-State High Plains–Ogallala Aquifer Regional Resources Study. Austin, Texas: High Plains Associates, 1982.

Soil Conservation in America: What Do We Have to Lose? Washington, D.C.: American Farmland Trust, 1984.

Taylor, James H., Eddie C. Burt, and Alvin Bailey. Tire Options and Consequences for Four-Wheel-Drive Tractors. SAE Technical Paper Series. Warrendale, Pennsylvania: Society of Automotive Engineers, 1979.

A Time to Choose: Summary Report on the Structure of Agriculture. Washington, D.C.: U.S. Department of Agriculture, 1981.

U.S. Bureau of the Census. Historical Statistics of the United States, Colonial Times to 1970, Bicentennial Edition, Part I. Washington, D.C.: U.S. Government Printing Office, 1975.

U.S. Comptroller General. Financial Condition of American Agriculture. Washington, D.C.: General Accounting Office, 1985.

U.S. Congress, Congressional Budget Office. Public Policy and the Changing Structure of American Agriculture. Washington, D.C.: U.S. Government Printing Office, 1978.

*U.S. Congress, Office of Technology Assessment. Technology, Public Policy, and the Changing Structure of American Agriculture. Washington, D.C.: U.S. Government Printing Office, 1986. Though much too long and wordy, a valuable study of the effect of new technologies on agriculture nevertheless.

U.S. Congress, Office of Technology Assessment. Water-Related Technologies for Sustainable Agriculture in the U.S. Arid/Semiarid Lands. Washington, D.C.: U.S. Government Printing Office, 1983.

U.S. Congress, Senate, Select Committee on Small Business. Ownership and Control of Farmland in the United States. Committee Print. Washington, D.C.: U.S. Government Printing Office, 1980.

USDA Study Team on Organic Farming. Report and Recommendations on Organic Farming. Washington, D.C.: U.S. Department of Agriculture, 1980.

U.S. Department of Agriculture. 1985 Agricultural Chartbook. Washington, D.C.: U.S. Government Printing Office, 1985.

U.S. Department of Agriculture. 1986 Fact Book of Agriculture. Washington, D.C.: U.S. Government Printing Office, 1985.

U.S. President's Council on Environmental Quality. Environmental Quality: The Sixth Annual Report. Washington, D.C.: U.S. Government Printing Office, 1975.

Veseth, Roger, et al. *Fertilizer Band Location for Cereal Root Access.* Crop Management Series, No-Till and Minimum Tillage Farming. A Pacific Northwest Extension Publication. Pullman, Washington, Moscow, Idaho, Corvallis, Oregon: USDA Cooperative Extension Service, 1986.

Whitehead, Vivian B. *A List of References for the History of Agricultural Technology.* Davis, California: Agricultural History Center, University of California, 1979.

Womach, Jasper. *Section 22 Authority: Protecting USDA Price Support Programs from Competitive Imports.* Washington, D.C.: Congressional Research Service, Library of Congress, 1984.

Unpublished Manuscripts

Church, Lillian. *History of the Plow.* Washington, D.C: U.S. Department of Agriculture, Bureau of Agricultural Engineering, 1935.

Madden, Patrick. *Case Studies of Organizational Linkages and Technology Transfer,* Volume 4 of *The Agricultural Technology Delivery System* by Irwin Feller, et al. University Park, Pennsylvania: Institute for Policy Research and Evaluation, Pennsylvania State University, 1984.

*Steiner, Frederick. *Soil Conservation: Politics, Policy, and Planning.* Pullman, Washington: Washington State University, Department of Horticulture and Landscape Architecture, to be published. Will be an excellent sourcebook for policy analysts.

Special Magazine and Journal Issues

*"Conservation Tillage: A Special Issue." *Journal of Soil and Water Conservation,* May–June 1983. A 190-page collection of papers and reports on conservation tillage. Meant primarily for a scientific-academic audience. Some individual papers are cited below.

Conservation Tillage Guide. A *Successful Farming* publication. Des Moines, Iowa: Meredith Corporation, 1983.

"Non-Point Source Pollution: A Special Issue." *Journal of Soil and Water Conservation,* January–February 1985.

Soil Sense '86. An annual published by California Farmer Publishing Company, 731 Market Street, San Francisco, California 94103. Contains useful listings of agribusinesses and university and government experts involved with conservation tillage. New editions are available in the fall of each year.

Articles

Bedunah, Don, and E. Earl Willard. "Sod Busting in Montana: The Taming of the West?" *Western Wildlands,* Spring 1984, pp. 2–7.

Breimyer, Harold. "On the Economy: The Cost of Erosion." *Land Stewardship Letter,* Autumn 1985, pp. 1, 11.

Brewster, David. "Historical Notes on Agricultural Structure" and "The Family Farm: A Changing Concept." In *Structure Issues of American Agriculture*. Agricultural Economics Report 438. Washington, D.C.: U.S. Department of Agricuture, 1979.

Brink, Wellington. "Big Hugh's New Science." *The Land*, Spring 1951, pp. 90–94.

Cacek, Terry. "Organic Farming: The Other Conservation Farming System." *Journal of Soil and Water Conservation*, November–December 1984, pp. 357–360.

Clark, Edwin H., II. "The Off-Site Costs of Soil Erosion." *Journal of Soil and Water Conservation*, January–February 1985, pp. 19–22.

Clifford, Harlan C. "A Step Away from the Dust Bowl." *Newsweek*, 4 June 1984, p. 15.

"Computers Predict Pollution Potential of Pesticides." *American Horticulturist*, November 1985, n.p.

"Conservation Tillage Acreage Pushing 100 Million Mark." *Conservation Tillage News*, February 1986, p. 1.

Cook, R. J. "Biological Control of Plant Pathogens: Theory to Application." *Phytopathology*, Vol. 75, No. 1 (1985), pp. 25–29.

Crittenden, Ann. "More and More Conglomerate Links in the U.S. Food Chain." *New York Times*, 1 February 1981, p. E-3.

Crosson, P. "Conservation Tillage: The Public Benefits." In *Conservation Tillage: Strategies for the Future*. Conference proceedings edited by Hal D. Heimstra and Jim W. Bauder. Fort Wayne, Indiana: Conservation Tillage Information Center, 1985.

Dillman, Don. "Factors Influencing the Adoption of No-Till Agriculture." In *Proceedings of the 1985 No-Till Farming Winter Crop Production Seminar*, edited by Dave Huggins. Spokane, Washington: Yielder Drill Company, 1985.

Dillman, Don, and John E. Carlson. "Influence of Absentee Landlords on Soil Erosion Control Practices." *Journal of Soil and Water Conservation*, January–February 1982, pp. 37–41.

"Drought: Will History Repeat?" *U.S. News and World Report*, 4 April 1977, pp. 48–51.

Easterbrook, Gregg. "Making Sense of Agriculture." *Atlantic Monthly*, July 1985, pp. 63–78.

Ebeling, Walter. "Agriculture in America: Roots." *Wilson Quarterly*, Summer 1981, pp. 111–119.

Ebenreck, Sara. "The Tillage Revolution." *American Land Forum*, Summer 1983, pp. 8–10.

Elkins, Charles B. "Modifying Plowpans to Improve Cotton Root Growth." In *1985 Beltwide Cotton Production Research Conference Proceedings*, n.pub., 1985.

Elkins, Charles B. "A Slit-Plant Tillage System." *Transactions of the American Society of Automotive Engineers*, 1983, pp. 710–712.

"Federal List of 403 Very Toxic Chemicals, With Effects on Human Health." *New York Times*, 20 November 1985, D-30.

"Figuring Conservation Tillage Economics Is an Individual Matter." *Conservation Tillage News*, May 1986, p. 1.

"First Plow Came from Saw Blade." In *The Agriculture of Ohio*, Bulletin 326 (July 1918), p. 225.

Foell, R. H. "What Conservation Tillage Means to Agribusiness." In *Conservation Tillage: Strategies for the Future*. Conference proceedings edited by Hal D. Heimstra and Jim W. Bauder. Fort Wayne, Indiana: Conservation Tillage Information Center, 1985.

Fream, William, and Roland Truslove. "Agriculture." *Encyclopaedia Britannica*, Vol. 1. Cambridge: Cambridge University Press, 1911.

*Fussell, G. E. "Ploughs and Ploughing Before 1800." *Agricultural History*, July 1966, pp. 177–186. History of agriculture from the traction plow of 4000 B.C. to the moldboard plow originating in the Low Countries and adopted and improved in England in the eighteenth century.

"Gearing Up for the Conservation Reserve." *Conservation Focus* (publication of the National Association of State Departments of Agriculture Research Foundation), February 1986, pp. 1–2.

"Greening the Gene." *Newsweek*, 12 November 1984, pp. 103–104.

"Groundwater Pollution Cited at Groundwater Protection Conference." *Alternative Agriculture News*, January 1986, n.p.

"Here's How Herman Warsaw Produced 370 Bushels per Acre Corn Yield." *Successful Farming*, January 1986, p. 37.

*Hinkle, Maureen K. "Problems with Conservation Tillage." *Journal of Soil and Water Conservation*, May–June 1983, pp. 201–206. Summary of environmental risks by a National Audubon Society analyst.

Kladivko, Eileen J., Alec D. Mackay, and Joe M. Bradford. "Earthworms as a Factor in the Reduction of Soil Crusting." *Soil Science Society of America Journal*, Vol. 50 (1986), pp. 191–196.

Langdale, G. W., and R. A. Leonard. "Nutrient and Sediment Losses Associated With Conventional and Reduced Tillage Agricultural Practice." In *Nutrient Cycling in Agricultural Ecosystems*, edited by R. Lawrance, et al. Watkinsville, Georgia: University of Georgia College of Agriculture Experiment Station, 1983.

Larson, W. E. "Tillage During the Past 25 Years." *Crops and Soils*, December 1972, pp. 5–6.

Lerza, Catherine. "Anybody Interested in the Future of the Family Farm?" *Rural Coalition Report*, July–September 1983, pp. 1–7.

Little, Charles E. "Needed: New Assumptions for Agriculture." *American Land Forum*, Spring 1986, p. 25.

Little, Charles E. "The No-Till Revolution." *Country Journal*, September 1985, pp. 54–61.

Lorang, Glenn. "Feed the Wheat, Starve the Weeds with Fertilizer Below the Seed." *Farm Journal*, November 1983, pp. 22–24.

Lord, Russell. "Chief of the Soil Revival." *The Land*, Spring 1951, pp. 83–89.

Mackay, Alec D., and Eileen J. Kladivko. "Earthworms and Rate of Breakdown of Soybean and Maize Residues in Soil." *Soil Biology Biochemistry*, Vol. 7, No. 6 (1985), pp. 851–857.

Mannering, Jerry, and Charles R. Fenster. "What Is Conservation Tillage?" *Journal of Soil and Water Conservation*, May–June 1983, pp. 141–143.

Massee, T. W. "Conservation Tillage Obstacles on Dryland." *Journal of Soil and Water Conservation*, July–August 1983, pp. 339–341.

Massee, T. W., and H. O. Waggoner. "Productivity Losses from Soil Erosion on Dry Cropland in the Intermountain Area." *Journal of Soil and Water Conservation*, September–October 1985, pp. 447–450.

Meyer, L. Donald, and William C. Moldenhauer. "Soil Erosion by Water: The Research Experience." *Agricultural History*, April 1985, pp. 192–204.

*Moldenhauer, W. C., et al. "Conservation Tillage For Erosion Control." *Journal of Soil and Water Conservation*, May–June 1983, pp. 144–151. Authoritative and admirably brief technical summary of the potentials of conservation tillage to reduce erosion in various parts of the country.

Myers, Peter C. "Why Conservation Tillage." *Journal of Soil and Water Conservation*, May–June, 1983, p. 136.

Odell, Rice. "Biotechnology: How Tight Must Our Control Be?" *Conservation Foundation Letter*, May–June 1985, pp. 1–6.

Palmer, Lane. "300-Bushel Corn: High Residue 'Crucial'." *Journal of Soil and Water Conservation*, May–June, 1983, p. 262.

*Papendick, R. I. "Changing Tillage and Cropping Systems: Impacts, Recent Developments, and Emerging Research Needs." In *The Optimum Tillage Challenge*, Proceedings of the Saskatchewan Institute of Agrologists, Update Series, edited by Glen Hass. Saskatoon, Saskatchewan: University of Saskatchewan Printing Services, 1984. A well-written roundup of no-till benefits and problems in the Northwest. For a photocopy, try Agricultural Research Center, Washington State University, Pullman, Washington 99164-6421.

"PIK: Multimillion-Dollar Crop Swap or Flop?" *Rural Coalition Report*, March 1984, pp. 6–7.

Pimintel, David, et al. "Land Degradation." *Science*, 8 October 1976, pp. 149–154.

Price, Vincent J. "Minimum Tillage: Looks Like a Winner." *Soil Conservation*, October 1972, pp. 43–44.

Risser, James. "The Other Farm Crisis." *Sierra*, May–June 1985, pp. 40–47.

*Rodale, Robert. "Breaking New Ground: The Search for Sustainable Agriculture." *The Futurist*, February 1983, pp. 16–20. The trouble with Jethro Tull and the need for "regenerative agriculture."

Rosen, Frederic A. "A Tough Row Not to Hoe." *American Way*, March 1984, pp. 68–70.

Sampson, R. Neil. "A Landmark for Soil Erosion." *American Land Forum*, Spring 1986, pp. 15–17.

Sampson, R. Neil. "What Must Be Done About Soil Erosion." *American Land Forum*, Spring 1982, pp. 9–13.

Samuelson, Robert J. "The Farm Mess Forever." *Newsweek*, 18 November 1985, p. 74.

Schmidt, William F. "South's Farm Loss Put at $1 Billion." *New York Times*, 23 July 1986, pp. A1, D21.

Schneider, Keith. "Cost of Farm Law Might Be Double Original Estimate." *New York Times*, 22 July 1986, pp. A1, A15.

Schneider, Keith. "Debt Notices Go to 65,000 Farmers." *New York Times*, 6 April 1986, p. A-28.

Schneider, Keith. "Farm Trade Deficit a Record in May." *New York Times*, 28 June 1986, pp. 35, 38.

Schneider, Keith. "Technology: Customized Crop Plants." *New York Times*, 29 May 1986, p. D-2.

Sinclair, Ward. "The World Doesn't Need Our Farmers." *Washington Post*, 29 December 1985, p. B-l.

"Slightly More Herbicide Used in Conservation Tillage, Lake Erie Study Finds." *Conservation Tillage News*, June 1984, p. 1.

"Take Lessons from the Father of No-Till." *Soil Sense '86*, pp. 16–20.

Taylor, James H. "Benefits of Permanent Traffic Lanes in a Controlled Traffic Crop Production System." *Soil and Tillage Research*, 3 (1983), pp. 385–395.

Taylor, James H. "A Controlled Traffic Agricultural System Using a Wide-Frame Carrier." In *Proceedings, 7th International Conference of Terrain-Vehicle Systems*. Calgary, Alberta: International Society for Terrain-Vehicle Systems, Inc., 1981.

Thompson, Dick and Sharon. "Cut Your Weed Control Costs in Half." *New Farm*, May–June 1985, pp. 24–27, 30–31.

Thompson, Dick and Sharon. "Family Planning—Farm Style." *New Farm*, May–June 1985, p. 48.

Thurlow, D. L., and C. B. Elkins. "Effect of In-Row Chisel at Planting on Yield and Growth of Full Season Soybeans." *Highlights of Agricultural Research* (Publication of Alabama Agricultural Experiment Station, Auburn University, Alabama), Winter 1983, n.p.

"Tillage Club Provides Support Base Farmers Need for Change." *Conservation Tillage News*, June 1986, p. 1.

Timmons, John F. "Protecting Agriculture's Natural Resource Base." *Journal of Soil and Water Conservation*, January–February 1980, pp. 5–11.

Triplett, Glover B., and David M. Van Doren, Jr. "Agriculture Without Tillage." *Scientific American*, January 1977, pp. 28–33.

Tugwell, Rexford Guy. "Down to Earth" (1936). In *American Habitat: A Historical Perspective*, edited by Barbara Gutmann Rosencrantz and William A. Koelch. New York: The Free Press, 1973.

"A Vermont Farm." *Fortune*, February 1939, pp. 48–52, 97–104.

Veseth, Roger. "The Unique Northwest Erosion Problem." *STEEP Extension Newsletter*, February–March 1986, pp. 8–11.

Walchek, Ken. "A 1980s Dustbowl?" *Montana Outdoors*, September–October 1983, pp. 15–19, 26.

Weller, D. M. "Suppression of Take-All of Wheat by Seed Treatment with Fluorescent Pseudomonads." *Phytopathology*, Vol. 73, No. 3 (1983), pp. 463–469.

Mimeographed or Photocopied Papers

Doyle, Jack. "The Intentional Release of Genetically Engineered Organisms." Washington, D.C.: Environmental Policy Institute, 1984. Testimony before the U.S. Senate Subcommittee on Toxic Substances and Environmental Oversight.

Doyle, Jack. "Testimony Before the Subcommittee on Patents, Copyrights and Trademarks." Washington, D.C.: Environmental Policy Institute, 1985. U.S. Senate Hearing, Washington, D.C.

Elkins, Charles B. "Slit Tillage." Auburn, Alabama: Auburn University, 1984.

Griffith, D. R., et al. "Maximizing Returns to Alternative Tillage Systems." West Lafayette, Indiana: Purdue University, n.d.

Griffith, Donald R., and Marvin L. Swearingin. "Basics for No-Tilling Soybeans." West Lafayette, Indiana: Purdue University, 1983. Paper for the Industrial Plant Food and Agricultural Chemical Conference.

Hallberg, George R. "Agrichemicals and Water Quality." Iowa City, Iowa: Iowa Geological Survey, 1986. Paper presented to the National Research Council Colloquium on Agrichemical Management to Protect Water Quality, Washington, D.C.

Kladivko, Eileen. "Effects of Tillage on Soil Biology." West Lafayette, Indiana: Purdue University, 1986. Paper for Indiana No-Till Conference.

*Riggleman, James D. "Future Priorities In Weed Science." E. I. DuPont de Nemours and Co., Inc., 1986. Presidential address before the Weed Science Society of America, 1986. A first-rate summary of weed scientists' agenda, herbicidal and otherwise. May be published later. For information or a photocopy, contact the author at DuPont's Agricultural Products Department, Barley Mill Plaza, Wilmington, Delaware 19898.

Thompson, Dick and Sharon. "Thompson's Regenerative Agriculture Demonstration Farm." Boone, Iowa, n.d.

Volak, Barney. "Conservation Tillage Projects." Kutztown, Pennsylvania: Rodale Research Center, n.d. (1985?)

Pamphlets, Brochures, and Audiovisual Material

The Bush Hog Ro-Till Solves Soil Compaction Problems. Selma, Alabama: Bush Hog Manufacturing Company, n.d.

Cain, Steve. *Farming in the Profit Zone.* Goodfield, Illinois: DMI, Inc., 1981.

Common Sense Conservation. Moline, Illinois: Deere & Company, n.d.

*Don't Let Soil Compaction Squeeze Your Profits. Indianapolis, Indiana: Elanco Products Company, n.d. An excellent illustrated summary of what compaction is and its effects on the soil. Available from Elanco at 740 S. Alabama St., Indianapolis, Indiana 46285.

Fine Tuning Your No-Till Techniques. San Francisco: Chevron Chemical Company, n.d.

Is PAT for You? Selma, Alabama: Bush Hog Manufacturing Company, n.d.

Modern Communications Group. "PAT: The Tillage Revolution." Mobile, Alabama, 1985. Videocassette recording.

Paraquat in Perspective. Wilmington, Delaware: ICI Americas, Inc., n.d. (1985?)

Powers, D. T., Geoff Luth, and Greg Schmick. "Yielder No-Till Grain Drill." Spokane, Washington: Yielder Drill Company, 1985. Videocassette recording.

Ridge Planting. St. Louis, Missouri: Monsanto Agricultural Products Company, 1983. A useful how-to and inspirational booklet containing reprints of articles from *Successful Farming Magazine.* Monsanto's address is 800 North Lindbergh Blvd., St. Louis, Missouri 63167.

Soybeans '86: Ideas for Profit. Research Triangle Park, North Carolina: Union Carbide Agricultural Products Company, Inc., n.d. (1985?)

Steps for Implementing a Conservation Tillage Demonstration Project. Fort Wayne, Indiana: Conservation Tillage Information Center, 1985.

Tillage for the Times. Mundelein, Illinois: International Minerals & Chemical Corporation, n.d. (1985?).

Conservation Tillage Information Center Survey of Tillage Practices, 1985

Excerpts from the Executive Summary

Since 1982, the Conservation Tillage Information Center has conducted a survey of tillage practices to determine the extent of adoption of conservation tillage. These excerpts constitute only a very small fraction of the data available. Readers may order the full report which contains data by region, by crop, by type of tillage, plus state by state summaries. In addition, CTIC has single state reports with data down to the county level. The truly statistically inclined may wish to inquire about custom-tailored reports, or even get the raw data, available on diskette. The address is CTIC, 2010 Inwood Drive, Executive Park, Ft. Wayne, IN 46815. The phone number is (219) 426-6642.

□   □   □

1. DEFINITION OF CONSERVATION TILLAGE

Any tillage and planting system that maintains at least 30 percent of the soil surface covered by residue after planting to reduce soil erosion by water; or, where soil erosion by wind is the primary concern, maintains at least 1000 pounds of flat, small-grain residue equivalent on the surface during the critical erosion period.

2. TYPES OF CONSERVATION TILLAGE

A. No-Till

The soil is left undisturbed prior to planting. Planting is completed in a narrow seedbed approximately 1–3 inches wide. Weed control is accomplished primarily with herbicides.

B. Ridge-Till

The soil is left undisturbed prior to planting. Approximately one-third of the soil surface is tilled at planting with sweeps or row cleaners. Planting is completed on ridges usually 4–6 inches higher than the row middles. Weed control is accomplished with a combination of herbicides and cultivation. Cultivation is used to rebuild ridges.

C. Strip-Till

The soil is left undisturbed prior to planting. Approximately one-third of the soil surface is tilled at planting time. Tillage in the row may consist of a rototiller, in-row chisel, row cleaners, and so forth. Weed control is accomplished with a combination of herbicides and cultivation.

D. Mulch-Till

The total soil surface is disturbed by tillage prior to planting. Tillage tools such as chisels, field cultivators, discs, sweeps, or blades are used. Weed control is accomplished with a combination of herbicides and cultivation.

E. Reduced-Till

Any other tillage and planting system not covered above that meets the 30 percent residue requirement.

3. INTERPRETING THE DATA

The following items should be noted when reviewing the data on the tables following:

- □ The "Acres" column includes both conservation tillage and conventional tillage. "Acres" minus "Conservation Tillage" equals Conventional Tillage.

- □ The total of the "Acres" column may exceed the cropland base due to more than one crop being planted on the same acreage in 1985 through double cropping.

- □ Totals and percentage sums do not include permanent pasture, fallow, or conservation use. The percentages shown are based on the totals that do not include these categories.

- "Conservation Tillage" is the sum of "No-Till," "Ridge-Till," "Strip-Till," "Mulch-Till," and "Reduced-Till."

- "Small Grain" includes such crops as wheat, oats, and barley.

- "Vegetable and Truck Crops" includes commercially grown crops such as potatoes, tomatoes, and melons.

- "Other Crops" includes crops not listed in another category.

- "Fallow" includes cropland not planted, such as dryland summer fallow or idle land.

- "Conservation Use" includes cropland idled for government diversion programs such as PIK, set aside, or diverted acres.

4. NATIONAL DATA

Acres by Various Conservation Tillage Types

Category	Acres	No-Till	Ridge-Till	Strip-Till	Mulch-Till	Reduced Till	Conservation Tillage*
Corn (FS)	81,909,218	5,759,670	1,295,454	412,594	21,241,859	2,857,693	31,567,270
Corn (DC)	1,006,179	314,294	500	1,925	96,910	59,205	472,835
Small Grain	99,616,871	2,635,469	22,405	22,080	21,036,378	10,665,663	34,581,995
Soybeans (FS)	59,180,983	1,632,869	516,732	64,355	14,005,559	1,645,387	17,854,902
Soybeans (DC)	6,743,606	2,558,018	8,120	30,293	973,436	439,698	4,009,565
Cotton	11,073,084	8,179	7,517	7,023	198,660	116,315	337,694
Grain Sorghum (FS)	18,459,430	724,702	92,106	38,950	3,307,923	1,746,396	5,910,077
Grain Sorghum (DC)	1,289,391	282,505	2,120	5,272	262,873	137,104	669,874
Veg. and Truck Crops	6,033,288	34,657	11,539	8,986	177,564	109,119	341,865
Forage Crops	10,908,641	465,804	0	0	1,147,177	452,513	2,065,494
†Permanent Pasture	3,798,511	539,767	0	0	394,063	286,366	1,220,196
Other Crops	20,640,140	333,728	6,430	11,625	861,493	509,388	1,722,664
†Fallow	27,938,428	1,258,224	0	0	4,888,980	3,890,879	10,038,083
†Conservation Use	30,466,467	0	0	0	0	0	0
Totals	316,862,831	14,949,895	1,962,923	623,103	63,309,832	18,738,482	99,584,235

*Sum of No-Till, Ridge-Till, Strip-Till, Mulch-Till, and Reduced Till

†Not Included in Totals

FS = Full Season; DC = Double Crop

Percentages of Various Conservation Tillage Types

Category	No-Till	Ridge-Till	Strip-Till	Mulch-Till	Reduced Till	Conservation Tillage*
Corn (FS)	7.03	1.58	0.50	25.93	3.49	38.54
Corn (DC)	31.17	0.05	0.19	9.61	5.67	46.90
Small Grain	2.85	0.02	0.02	21.12	10.71	34.71
Soybeans (FS)	2.76	0.87	0.14	23.67	2.78	30.22
Soybeans (DC)	37.93	0.12	0.45	14.43	6.52	59.46
Cotton	0.07	0.07	0.06	1.79	1.05	3.05
Grain Sorghum (FS)	3.93	0.50	0.21	17.92	9.46	32.02
Grain Sorghum (DC)	21.91	0.16	0.41	20.39	10.63	53.50
Veg. and Truck Crops	0.57	0.19	0.15	2.94	1.81	5.67
Forage Crops	4.27	0.00	0.00	10.52	4.15	18.93
†Permanent Pasture	14.21	0.00	0.00	10.37	7.54	32.12
Other Crops	1.62	0.03	0.06	4.17	2.47	8.35
†Fallow	4.50	0.00	0.00	17.50	13.93	35.93
†Conservation Use	0.00	0.00	0.00	0.00	0.00	0.00
Totals	4.72	0.62	0.20	19.95	5.91	31.43

*Sum of No-Till, Ridge-Till, Strip-Till, Mulch-Till, and Reduced Till
†Not Included in Totals
FS = Full Season; DC = Double Crop

5. STATE-BY-STATE DATA

Acres by Various Conservation Tillage Types

State	Acres	No-Till	Ridge-Till	Strip-Till	Mulch-Till	Reduced Till	Conservation Tillage*
Alabama	3,451,757	261,096	60	66,600	470,365	0	798,121
Alaska	33,415	18	0	0	6,151	540	6,707
Arizona	993,905	4,820	0	0	16,675	34,104	55,599
Arkansas	6,958,143	110,074	1,380	50	247,388	516,246	875,140
California	7,882,582	40,788	1,310	50	61,967	154,175	256,290
Colorado	7,187,727	154,102	27,248	32,440	1,074,289	1,320,862	2,608,941
Connecticut	89,834	6,844	45	0	6,300	0	13,189
Delaware	616,856	256,182	0	700	220,519	0	477,401
Florida	2,614,482	39,669	9,725	543	56,630	109,714	215,281
Georgia	5,888,599	246,396	450	7,057	1,110,408	5,500	1,369,811
Hawaii	316,850	0	0	0	0	4,500	4,500
Idaho	4,508,943	209,524	1,036	0	488,290	384,522	1,083,372
Illinois	23,283,444	1,461,697	202,330	12,640	6,079,901	0	7,756,568
Indiana	12,425,826	926,959	171,824	22,035	4,011,333	19,500	5,151,651
Iowa	23,705,631	822,168	277,908	48,811	10,927,865	0	12,076,752
Kansas	21,828,162	659,758	66,790	5,940	5,366,115	5,082,772	11,181,375
Kentucky	4,294,298	962,112	600	255	452,589	922,980	2,338,536
Louisiana	4,628,984	51,683	0	45	91,605	123,449	266,782
Maine	310,885	1,901	600	0	28,505	25,147	56,153
Maryland	1,526,346	462,546	0	1,100	587,838	0	1,051,484
Massachusetts	79,536	2,110	0	0	569	1,575	4,254
Michigan	6,555,558	325,361	16,271	3,876	1,379,881	0	1,725,389
Minnesota	19,721,833	125,115	384,341	29,792	3,827,168	10,875	4,377,291
Mississippi	5,415,054	120,119	1,086	10	212,705	450,086	794,006
Missouri	11,402,711	896,891	9,005	11,441	3,493,092	0	4,410,429
Montana	8,545,014	371,633	5,200	162	1,548,544	1,418,859	3,344,598

166

Nebraska	16,360,355	531,469	490,144	275,413	3,936,189	1,302,457	6,535,672
Nevada	142,523	4,243	0	0	25,342	33,283	62,868
New Hampshire	28,526	1,279	0	0	1,258	1,533	4,070
New Jersey	521,588	80,001	0	0	205	69,278	149,484
New Mexico	1,325,346	24,695	6,340	2,650	179,824	285,937	499,449
New York	2,733,676	69,624	518	400	59,845	213,768	344,155
North Carolina	5,177,722	705,942	39,093	19,115	442,911	317,182	1,524,243
North Dakota	19,058,627	653,063	13,070	0	1,910,497	1,872,444	4,,449,074
Ohio	9,954,194	1,263,351	47,400	500	1,897,540	514,110	3,722,901
Oklahoma	9,673,761	80,319	5,190	12,180	2,275,147	495,330	2,868,166
Oregon	2,283,742	130,168	0	440	314,242	223,310	668,160
Pennsylvania	3,149,644	610,931	1,006	50	436,261	300,464	1,348,712
Rhode Island	9,255	468	0	0	280	0	748
South Carolina	2,884,847	130,781	150	9,915	37,416	82,233	260,495
South Dakota	13,896,920	342,313	83,057	10,401	4,456,885	0	4,892,656
Tennessee	4,006,240	450,992	2,520	1,454	528,136	1,335	984,437
Texas	24,841,453	307,824	45,011	19,870	2,140,582	1,589,863	4,103,150
Utah	654,858	27,207	0	0	28,646	46,885	102,738
Vermont	145,253	4,547	0	0	250	2,344	7,141
Virginia	2,357,711	691,632	3,000	6,138	349,782	251,834	1,302,386
Washington	5,354,466	95,321	10,000	4,000	1,226,848	282,263	1,618,432
West Virginia	151,319	49,026	0	0	13,092	9,114	71,232
Wisconsin	5,962,516	166,620	38,855	17,030	1,194,568	195,067	1,615,140
Wyoming	813,860	6,704	360	0	60,095	38,464	105,623
Puerto Rico	76,024	1,606	0	0	27,299	11,576	40,463
National	316,862,831	14,848,885	1,962,923	623,103	63,309,832	18,738,482	99,584,235

*Sum of No-Till, Ridge-Till, Strip-Till, Mulch-Till, and Reduced Till

II

National and State Sources of Information on Conservation Tillage

1. NATIONAL

NATIONAL ASSOCIATION OF CONSERVATION DISTRICTS

NACD represents conservation districts and promotes soil conservation programs throughout the country. NACD uses the Conservation Tillage Information Center (CTIC) to provide information on adoption of conservation tillage practices.

James Lake
Director
Conservation Tillage Information Center
2010 Inwood Drive
Executive Park
Ft. Wayne, IN   46815
(219)426-6642

AGRICULTURAL RESEARCH SERVICE

A division of USDA, this federal agency conducts national research on soil and water conservation, crop and animal production and protection, and human nutrition.

James F. Power
Agricultural Research Service
Keim Hall, East Campus
University of Nebraska
Lincoln, NE   68585
(402)472-1484

SOIL CONSERVATION SOCIETY OF AMERICA

SCSA is dedicated to promoting the science and art of good land use. SCSA has over 13,000 members in the United States, Canada, and 80 other foreign countries.

Walter Peechatka
Executive Vice President
Soil Conservation Society of America
7515 NE Ankeny Rd.
Ankeny, IA   50021
(515)289-2331

RODALE INSTITUTE

The Rodale Institute is a nonprofit education and research organization promoting regenerative farming and economic development in the United States and in developing countries.

James Morgan
Executive Director
Rodale Institute
222 Main St.
Emmaus, PA   18049
(215)967-5171

INSTITUTE FOR ALTERNATIVE AGRICULTURE

The Institute offers scientific information on alternative agriculture to government agencies, nonprofit organizations, and individuals, and can put people in touch with technicians and scientists with conservation tillage experience.

Garth Youngberg
Executive Director
Institute for Alternative Agriculture
9200 Rd. #117
Greenbelt, MD   20770
(301)441-8777

AGRICULTURAL STABILIZATION AND CONSERVATION SERVICE

Part of USDA, ASCS is responsible for commodity production adjustment and support programs; national disaster assistance to agricultural producers through payments and cost-sharing; and certain national emergency-preparedness activities.

Vincent E. Gerimes
Branch Chief
ASCS-USDA
P.O. Box 2415
Washington, DC   20013
(202)447-6621

SOIL CONSERVATION SERVICE

Part of USDA, SCS provides technical assistance on conservation tillage and other soil and water conservation practices for individuals, groups, organizations, cities, towns, and county and state governments.

David Shertz
National Agronomist
USDA-SCS
P.O. Box 2890
Washington, DC   20013
(202)447-3783

2. STATE-BY-STATE

The following is a state-by-state breakdown of USDA Conservation Agronomists and Extension Service representatives. The role of Conservation Agronomists is to examine the effect soil fertilizers, insecticides, crop productivity, and other conservation tillage practices has in the overall conservation of the land. The role of the Extension Service representatives is to help individuals and groups make management decisions for use of renewable and non-renewable resources while assuring protection of the environment.

Conservation Agronomist	*Extension Service Representative*
Alabama	
Ernest Todd	Louis Chapman
665 Opelika Road	Auburn University
P.O. Box 311	Extension Hall
Auburn, AL 36830	Auburn, AL 36849
(205)821-8070	(205)826-4985
Alaska	
Burton Clifford	Donald Quarberg
201 East 9th Ave. #300	University of Alaska
Anchorage, AK 99501	Box 349
(907)271-2424	Delta Junction, AK 99737
	(907)895-4215

Conservation Agronomist	*Extension Service Representative*

Arizona

Verne Bathurst
201 E. Indianola Ave. #200
Phoenix, AZ   85012
(602)241-2247

Wilford Gardner
University of Arizona
Soils Dept.
Tucson, AZ   85721
(602)621-7228

Arkansas

Albert Sullivan
700 West Capitol Ave. #5423
Little Rock, AR   72201
(501)378-5445

Woody Miley
University of Arkansas
P.O. Box 391
Little Rock, AR   72203
(501)373-2500

California

Eugene Andreuccetti
2121C Second St. #102
Davis, CA   95616
(916)449-2848

Ronald Meyer
University of California
129 Hoagland
Davis, CA   95616
(916)752-2531

Colorado

Sheldon Boone
2490 West 26th Ave.
Diamond Hill, Bldg. A
Denver, CO   80211
(303)964-0292

Roy Follett
Colorado State University
Dept. of Agronomy
Ft. Collins, CO   80523
(303)491-6201

Connecticut

Philip Christensen
16 Professional Park Rd.
Storrs, CT   06268
(203)487-4011

Robert Peters
University of Connecticut
Plant Dept. SCI U-102
Storrs, CT   06268
(203)486-2928

Delaware

Douglas Hawkins
9 East Loockerman Street
Treadway Towers #207
Dover, DE   19901
(302)678-0750

Richard Fowler
University of Delaware
Townsend Hall
Newark, DE   19717
(302)451-2531

Florida

James Mitchell
401 S.E. 1st Ave.
Federal Bldg. #248
Gainesville, FL   32601
(904)377-0946

Raymond Gallaher
University of Florida
631 Wallace Bldg.
Gainesville, FL   32611
(904)392-2325

Georgia

Clayton Graham
355 East Hancock Ave.
Federal Bldg. Box 13
Athens, GA   30601
(404)546-2273

Owen Plank
University of Georgia
Agronomy Dept.
Athens, GA   30602
(404)542-9072

Conservation Agronomist	*Extension Service Representative*

Hawaii

Richard Duncan
300 Ala Moana Blvd.
P.O. Box 50004
Honolulu, HI   96850
(808)546-3165

Wade McCall
University of Hawaii
1910 East West Rd.
Honolulu, HI   96822
(808)948-8901

Idaho

Stanley Hobson
304 North 8th St. #345
Boise, ID   83702
(208)334-1601

Robert McDole
University of Idaho
Plant and Soil Dept.
Moscow, ID   83843
(208)885-6767

Illinois

John Eckes
301 North Randolph St.
Springer Federal Bldg.
Champaign, IL   61820
(217)398-5267

Robert Walker
University of Illinois
1301 W. Gregory Dr.
Urbana, IL   61801
(217)333-1130

Indiana

Robert Eddleman
5610 Crawfordsville Rd.
Corporate Square-West #2200
Indianapolis, IN   46224
(317)248-4350

Donald Griffith
Purdue University
Lilly Hall #2-416
West Lafayette, IN   47907
(317)494-4798

Iowa

Michael Nethery
210 Walnut St.
693 Federal Bldg.
Des Moines, IA   50309
(515)284-4261

Minoru Amemiya
Iowa State University
Dept. of Agronomy
Ames, IA   50011
(515)294-1923

Kansas

James Habiger
760 South Broadway
Salina, KS   67401
(913)823-4565

John Hickman
Kansas State University
Throckmorton Hall
Manhattan, KS   66506
(913)532-5776

Kentucky

Randall Giessler
333 Waller Ave. #305
Lexington, KY   40504
(606)233-2749

Kenneth Wells
University of Kentucky
N-122 Agriculture Science
Lexington, KY   40546
(606)257-4768

Louisiana

Horace Austin
3737 Government St.
Alexandria, LA   71302
(318)473-7751

Thomas Burch
Louisiana State University
270 Knapps Hall
Baton Rouge, LA   70803
(504)388-2186

Conservation Agronomist	Extension Service Representative

Maine

Ronald Hendricks
University of Maine
USDA Bldg.
Orono, ME   04473
(207)866-2132

Lewis Wyman
University of Maine
100 Winslow
Orono, ME   04469
(207)581-2940

Maryland

Pearlie Reed
4321 Hartwick Road
Hartwick Bldg. #522
College Park, MD   20740
(301)344-4180

Richard Weismiller
University of Maryland
Agromos Rd.
College Park, MD   20742
(301)454-4787

Massachusetts

Rex Tracy
451 West St.
Amherst, MA   01002
(413)256-0441

Stephen Herbert
University of Massachusetts
Stockbridge #126
Amherst, MA   01003
(413)545-2349

Michigan

Homer Hilner
1405 South Harrison Rd. #101
East Lansing, MI   48823
(517)337-6702

Frank Brewer
Michigan State University
11 Agricultural Hall
East Lansing, MI   48824
(517)355-0212

Minnesota

Donald Ferren
316 North Robert St.
200 Federal Bldg., Courthouse
St. Paul, MN   55101
(612)725-7675

John Moncrief
University of Minnesota
Dept. of Soil Science
St. Paul,, MN   55108
(612)373-1060

Mississippi

Vacant
100 West Capitol St.
Federal Bldg. #1321
Jackson, MS   39269
(601)965-5205

Duane Tucker
Mississippi State University
Box 5446
Mississippi, MS   39762
(601)325-3430

Missouri

Paul Larson
555 Vandiver Drive
Columbia, MO   65202
(314)875-5214

Donald Pfiest
University of Missouri
200 Ag. Engineer
Columbia, MO   65211
(314)882-2731

Montana

Glen Loomis
10 East Babcock St.
Federal Bldg. #443
Bozeman, MT   59715
(406)587-6813

James Bauder
University of Montana
806 Leon Johnson Hall
Bozeman, MT   59717
(406)994-3515

Conservation Agronomist	*Extension Service Representative*

Nebraska

Kenton Inglis
100 Centennial Mall North
Federal Bldg. #345
Lincoln, NE 68508
(402)471-5300

Elbert Dickey
Nebraska State University
249 Chase Hall, E. Campus
Lincoln, NE 68583
(402)472-3950

Nevada

Charles Adams
1201 Terminal Way #219
Reno, NV 89502
(702)784-5863

Clark Leedy
University of Nevada
Plant Science
Reno, NV 89557
(702)784-6981

New Hampshire

David Mussulman
Federal Bldg.
Durham, NH 03824
(603)868-7581

James Mitchell
University of New Hampshire
Plant Science
Durham, NH 03824
(603)862-3204

New Jersey

Joseph Branco
1370 Hamilton St.
Somerset, NJ 08873
(201)246-1662

James Justin
Rutgers University
Box 231 Lipman
New Brunswick, NJ 08903
(201)932-9872

New Mexico

Ray Margo, Jr.
517 Gold Ave., S.W. #3301
Albuquerque, NM 87102
(505)766-2173

Richard Baker
New Mexico State University
Box 77, Star Route
Clovis, NM 88101
(505)985-2292

New York

Dodd Paul
100 S. Clinton St.
James Hanley Federal Bldg. #771
Syracuse, NY 13260
(315)423-5521

Shaw Reid
154 Emerson
Cornell University
Ithaca, NY 14853
(607)256-2177

North Carolina

Bobbye Jack Jones
310 New Bern Ave.
Federal Office Bldg. #535
Raleigh, NC 27601
(919)856-4210

Jack Baird
1225 Williams
North Carolina State University
Raleigh, NC 27650
(919)373-3285

North Dakota

August Dornbusch, Jr.
Rosser Ave. and Third St.
P.O. Box 1458
Bismarck, ND 58502
(701)255-4011

Duane Berglund
108 Morrill #5655
North Dakota State University
Fargo, ND 58105
(701)237-8884

Conservation Agronomist	Extension Service Representative

Ohio

Harry Oneth
200 North High St. #522
Columbus, OH   43215
(614)469-6962

Donald Eckert
2021 Coffy Rd.
Ohio State University
Columbus, OH   43210
(614)422-2047

Oklahoma

Roland Willis
USDA Ag. Center Bldg.
Stillwater, OK   74074
(405)624-4360

James Steigler
245 Agriculture Hall
Oklahoma State University
Stillwater, OK   74078
(405)624-6422

Oregon

Jack Kanalz
1220 S.W. Third Ave.
Federal Bldg. #1640
Portland, OR   97204
(503)221-2751

James Vomocil
Soil Dept.
Oregon State University
Corvallis, OR   97331
(503)754-2441

Pennsylvania

James Olson
228 Walnut St.
Box 985, Federal Sq. Station
Harrisburg, PA   17108
(717)782-2202

Lynn Hoffman
119 Tyson
Pennsylvania State University
University Park, PA   16802
(814)692-7955

Rhode Island

Robert Klumpe
46 Quaker Lane
West Warwick, RI   02893
(401)828-1300

Michael Sullivan
Plant Science
University of Rhode Island
Kingston, RI   02881
(401)792-4540

South Carolina

Billy Abercrombie
1835 Assembly St.
Strom Thurmond Federal Bldg. #950
Columbia, SC   29201
(803)765-5681

Cliff Nolan
Agronomy Hall
University of South Carolina
Clemson, SC   29631
(803)656-3102

South Dakota

C. Budd Fountain
200 4th St., S.W.
Federal Building
Huron, SD   57350
(605)353-1783

Robert Durland
Ag. Engineer
South Dakota State University
Brookings, SD   57007
(605)688-5141

Tennessee

Donald Bivens
801 Broadway
675 Estes Kefauver, FB-USCH
Nashville, TN   37203
(615)736-5471

George Buntley
Box 1071
University of Tennessee
Knoxville, TN   37901
(615)974-7421

Conservation Agronomist	Extension Service Representative

Texas

Coy Garrett
101 S. Main Street
W.R. Poage Federal Bldg.
Temple, TX   76501
(817)774-1214

Edwin Colburn
Soil and Crop Management
Texas A&M University
College Station, TX   77843
(407)845-7967

Utah

Francis Holt
125 So. State St.
Federal Bldg. #4102
Salt Lake City, UT   84138
(801)524-5050

Philip Rasmussen
UMC/48
Utah State University
Logan, UT   84322
(801)750-2257

Vermont

John Titchner
69 Union St.
Winooski, VT   05404
(802)951-6795

Bill Jokela
Plant and Soil Dept.
University of Vermont
Burlington, VT   05405
(802)656-0480

Virginia

George Norris
400 North 8th St.
Federal Bldg. #9201
Richmond, VA   23240
(804)771-2455

Harlan White
419 Smyth Hall
Virginia Polytechnic Institute
Blacksburg, VA   24061
(703)961-6486

Washington

Lynn Brown
West 920 Riverside Ave. #360
Spokane, WA   99201
(509)456-3711

Carl Engle
Dept. of Ag. and Soil
Washington State University
Pullman, WA   99164
(509)335-2887

West Virginia

Rollin Swank
75 High St. #301
Morgantown, WV   26505
(304)291-4151

Charles Sperow
1076 Ag Science, Box 6108
University of West Virginia
Morgantown, WV   26506
(304)293-2219

Wisconsin

Cliffton Maguire
4601 Hammersley Rd.
Madison, WI   53711
(608)264-5577

Tommy Daniel
1585 Observation Drive
University of Wisconsin
Madison, WI   53706
(608)262-9969

Wyoming

Frank Dickson, Jr.
100 East "B" St.
Federal Office Bldg. #3124
Casper, WY   82601
(307)261-5201

Michael McNamee
Box 3354
University of Wyoming
Laramee, WY   82071
(307)766-3323

Index

Also Available from Island Press

Hazardous Waste Management: Reducing the Risk
By Benjamin A. Goldman, James A. Hulme, and Cameron Johnson
for the Council on Economic Priorities

Hazardous Waste Management: Reducing the Risk is a comprehensive sourcebook of facts and strategies which provides the analytic tools needed by policy makers, regulating agencies, hazardous waste generators, and host communities to compare facilities on the basis of site, management, and technology. The Council on Economic Priorities' innovative ranking system applies to real-world, site-specific evaluations, establishes a consistent protocol for multiple applications, assesses relative benefits and risks, and evaluates and ranks ten active facilities and eight leading commercial management corporations.

1986. xx, 316 pp., notes, tables, glossary, index.
Cloth, ISBN 0-933280-30-0. **$64.95**
Paper, ISBN 0-933280-31-9. **$34.95**

An Environmental Agenda for the Future
By Leaders of America's Foremost Environmental Organizations

". . . a substantive book addressing the most serious questions about the future of our resources."—John Chafee, Senator, Environmental & Public Works Committee. "While I am not in agreement with many of the positions the authors take, I believe this book can be the basis for constructive dialogue with industry representatives seeking solutions to environmental problems."—Louis Fernandez, Chairman of the Board, Monsanto Corporation.

The chief executive officers of the ten major environmental and conservation organizations launched a joint venture to examine goals the environmental movement should pursue now and on into the 21st century. This book presents policy recommendations to effect changes needed to bring about a healthier, safer living experience. Issues discussed include: nuclear issues, human population growth, energy strategies, toxic and pollution control, and urban environments.

1985. viii, 155 pp., bibliography.
Paper, ISBN 0-933280-29-7. **$5.95**

Land-Saving Action
Edited by Russell L. Brenneman and Sarah M. Bates

Land-Saving Action is the definitive guide for conservation practitioners. A written symposium by the 29 leading experts in land conservation, this book presents in detail land-saving tools and techniques that have been perfected by individuals and organizations across the nation. This is the first time such information has been available in one volume.

1984. xvi, 249 pp., tables, notes, author biographies, selected readings, index.
Cloth, ISBN 0-933280-23-8. **$39.95**
Paper, ISBN 0-933280-22-X. **$24.95**

Building an Ark: Tools for the Preservation of Natural Diversity through Land Protection
By Phillip M. Hoose

The author, The Nature Conservancy's national protection planner, presents a comprehensive plan to identify and protect each state's natural ecological diversity, and shows how plant and animal species can be saved from destruction without penalty to the landowner. Case studies augment this blueprint for conservation.

1981. xvi, 222 pp., key contacts: agencies, programs, legal experts, notes, tables, illustrations, bibliography, index.
Paper, ISBN 0-933280-09-2. **$12.00**

The Conservation Easement in California
By Thomas S. Barrett and Putnam Livermore for The Trust for Public Land

This is the authoritative legal handbook on conservation easements. This book examines the California law as a model for the nation. It emphasizes the effectiveness and flexibility of the California code. Also covered are the historical and legal backgrounds of easement technology, the state and federal tax implications, and solutions to the most difficult drafting problems.

1983. xiv, 173 pp., appendices, notes, selected bibliography, index.
Cloth, ISBN 0-933280-20-3. **$34.95**

Private Options: Tools and Concepts for Land Conservation
By Montana Land Reliance and Land Trust Exchange

Techniques and strategies for saving the family farm are presented by 30 experts. *Private Options* details the proceedings of a national conference and brings together, for the first time, the experience and advice of land conservation experts from all over the nation.

1982. xiv, 292 pp., key contacts: resource for local conservation organizations, conference participants, bibliography, index.
Paper, ISBN 0-933280-15-7. **$25.00**

Community Open Spaces
By Mark Francis, Lisa Cashdan, Lynn Paxson

Over the past decade, thousands of community gardens and parks have been developed on vacant neighborhood land in America's major cities. *Community Open Spaces* documents this movement in the U.S. and Europe, explaining how planners, public officials, and local residents can work in their own community to successfully develop open space.

1984. xiv, 250 pp., key contacts: resource organizations, appendices, bibliography, index.
Cloth, ISBN 0-933280-27-0. **$24.95**

Water in the West
By Western Network

An essential reference tool for water managers, public officials, farmers, attorneys, industry officials, and students and professors attempting to understand the competing pressures on our most important natural resource: water. This three-volume series provides an in-depth analysis of the effects of energy development, Indian rights, and urban growth on other water users.

1983. Vol. I: **What Indian Water Means to the West**
iv, 153 pp., key contacts: federal, state, and Indian agencies, maps, charts, documents, bibliography.
Paper, **$15.00**

Vol. II: **Water for the Energy Market**
   v, 162 pp., key contacts: federal, state, and Indian
   agencies, maps, charts, documents, bibliography,
   index.
   Paper, **$15.00**
Vol. III: **Western Water Flows to the Cities**
   v, 217 pp., maps, table of cases, documents, bibliography, index.
   Paper, **$25.00**

Green Fields Forever: The Conservation Tillage Revolution in America
By Charles E. Little

"*Green Fields Forever* is a fascinating and lively account of one of
the most important technological developments in American agriculture. . . . Be prepared to enjoy an exceptionally well-told tale,
full of stubborn inventors, forgotten pioneers, enterprising farmers—and no small amount of controversy."—Kenneth A. Cook, Sen-
ior Associate, World Wildlife Fund/Conservation Foundation.

1987. xx, 193 pp., illustrations, appendices, index, bibliography.
Cloth. ISBN 0-933280-35-1. **$24.95**
Paper. ISBN 0-933280-34-3. **$14.95**

Federal Lands: A Guide to Planning, Management, and State Revenues
By Sally K. Fairfax and Carolyn E. Yale

"In most of the western states, natural resource revenues are
extremely important as well as widely misunderstood. This book
helps to clarify states' dependencies on these revenues, which in
some instances may be near-fatal."—Don Snow, Director, Northern
Lights Institute. "An invaluable tool for state land managers. Here,
in summary, is everything that one needs to know about federal
resource management policies."—Rowena Rogers, President, Colorado State Board of Land Commissioners.

1987. xx, 252 pp., charts, maps, bibliography, index.
Paper. ISBN 0-933280-33-5. **$19.95**

Public Opinion Polling: A Handbook for Public Interest and Citizen Advocacy Groups
By Celinda Lake with Pat Callbeck Harper

"Lake has taken the complex science of polling and written a very
usable 'how-to' book. I would recommend this book to both can-

didates and organizations interested in professional, low-budget, in-house polling."—Stephanie Solien, Executive Director, Women's Campaign Fund. "This book should not be on the bookshelves, but in the hands of anyone who wants to conduct polls or is a consumer of public opinion polls."—Barbara G. Farah, Director of Surveys, *The New York Times.*

1987. 208 pp., tables, bibliography, appendix, index.
Paper. ISBN 0-933280-32-7. **$19.95**

These books are available directly from Island Press, Order Department, Box 7, Covelo, CA 95428. Please enclose $1.50 with each order for postage and handling; California residents add 6 percent sales tax. A catalog of current and forthcoming titles is available free of charge.

Island Press is a nonprofit organization dedicated to the publication of books for professionals and concerned citizens on the conservation and management of natural resources and the environment.

Photo Credits

Page 61. Tim McCabe. USDA—Soil Conservation Service.
Page 64. N. Cole. USDA—Soil Conservation Service.
Page 66. H. E. (Gene) Alexander. USDA—Soil Conservation Service.
Page 68. H. E. (Gene) Alexander. USDA—Soil Conservation Service.
Page 71. Fleischer Manufacturing, Inc.
Page 73. Fleischer Manufacturnig, Inc.
Page 75. Courtesy Carl and Rosemary Eppley.
Page 80. Tim McCabe. USDA—Soil Conservation Service.
Page 81. USDA—Agricultural Research Service.
Page 82. B. W. Anderson. USDA—Soil Conservation Service.
Page 85. Erwin W. Cole. USDA—Soil Conservation Service.
Page 87. Bush Hog Manufacturing Company.
Page 89. Bush Hog Manufacturing Company.
Page 90. USDA—Agricultural Research Service.
Page 92. USDA—Agricultural Research Service.
Page 94. USDA—Agricultural Research Service.
Page 96. Bush Hog Manufacturing Company.
Page 100. John Deere & Company.
Page 102. USDA—Agricultural Research Service.
Page 102. Tim McCabe. USDA—Soil Conservation Service.
Page 105. Harris & Ewing. Courtesy American Forestry Association.
Page 107. Robert C. Bjork. USDA—Office of Information.
Page 110. O. K. Blake. USDA—Soil Conservation Service.
Page 112. USDA—Agricultural Research Service.
Page 115. H. E. (Gene) Alexander. USDA—Soil Conservation Service.
Page 116. Rodale Press, Inc.
Page 119. Rodale Press, Inc.
Page 122. H. E. (Gene) Alexander. USDA—Soil Conservation Service.
Page 124. John Deere & Company.
Page 126. Lowary. USDA—Soil Conservation Service.
Page 128. R. S. Kellogg.
Page 128. U.S. Forest Service.
Page 130. Doug Wilson. USDA—Soil Conservation Service.
Page 133. USDA—Office of Information.
Page 135. USDA—Agricultural Research Service.
Page 138. Kurt Blum.
Page 140. Tim McCabe. USDA—Soil Conservation Service.

Production Management and Text Design   Greg Hubit Bookworks
Copyediting   Patricia Harris
Proofreading   Dorothy Wilson
Indexing   Jerilyn Emori
Composition   Graphic Typesetting Service
Printing and Binding   BookCrafters